AT THE PRECIPICE

AT THE PRECIPICE

NEW MEXICO'S CHANGING CLIMATE

LAURA PASKUS

UNIVERSITY OF NEW MEXICO PRESS | ALBUQUERQUE

Library of Congress Cataloging-in-Publication Data
Names: Paskus, Laura, 1974– author.
Title: At the precipice: New Mexico's changing climate / Laura Paskus.
Description: Albuquerque: University of New Mexico Press, 2020. |
    Includes bibliographical references and index.
Identifiers: LCCN 2020014478 (print) | LCCN 2020014479 (e-book) |
    ISBN 9780826359117 (paperback) |
    ISBN 9780826359124 (e-book)
Subjects: LCSH: Climatic changes—New Mexico. | Global warming—
    New Mexico. | Water-supply—Climatic factors—New Mexico.
Classification: LCC QC903.2. U6 P37 2020 (print) | LCC QC903.2. U6 (e-book) |
    DDC 363.7387409789—dc23
LC record available at https://lccn.loc.gov/2020014478
LC e-book record available at https://lccn.loc.gov/2020014479

Cover photograph by Pysberartist, licensed under CC by 2.0
Designed by Felicia Cedillos
Composed in Sabon LT Std 10/14.25

For Lillie, of course

Contents

Illustrations

Preface

Reporting on climate change in an arid region doesn't generate the excitement it might in states hammered by sea-level rise or hurricanes that knock out power, devastate cities, and cost many lives and billions of dollars. Here in the Southwest, forest die-offs happen over years and decades, sometimes in places people don't live or visit on a regular basis. When the Middle Rio Grande south of Albuquerque dries, as it does most summers—and as it did dramatically in the spring of 2018—the sandy channel is hidden for miles and miles behind walls of nearly impenetrable (and invasive) tamarisk and Russian olive. Drought is a slow burn. And when our wishes come true—when snowpack is robust or monsoon season delivers rains—we breathe a sigh of relief. We shrug off reports that average temperatures keep rising or that reservoirs still aren't anywhere near full. Because, who knows? Next year might be better.

There's another reason why reporting on climate change in a state like New Mexico can be so tricky: We have so much else to worry about. Whether it's the economy and a state budget lashed to the fossil-fuel industry or the state's chronic problems with educational funding, public and behavioral health, criminal justice, and poverty, people have a lot more to worry about than what might happen in a century.

But climate change is happening right now—and it will exacerbate all the other problems we grapple with here in the state.

Like earlier regional, national, and international reports on climate change, the Fourth National Climate Assessment, released in 2018, noted that the risks associated with climate change are highest for those who are already vulnerable, including low-income communities, some communities of color, children, and the elderly. The risks are particularly pronounced for tribal nations, which the US government restricted to the "driest portions of their traditional homelands." According to the report, the well-being of southwestern tribes is at increased risk from water scarcity, the loss of traditional foods, wildfire, and changes in the ranges of plants and animals.

Why the climate is changing, what its future impacts will be, and how those changes will affect humans, ecosystems, and the earth's systems are all things we already know. For some sixty years scientists and American political leaders have even known how to stop, or slow, warming. And in many ways New Mexico is a microcosm of what's happening across the globe politically, economically, and certainly when it comes to the impacts of climate change. Living within a "hotspot" of warming, New Mexicans are already watching how fossil-fuel extraction, "hot drought," and record-breaking annual temperature records play out in terms of forest fires, water challenges, and public-health impacts.

Time and again while reporting on climate change I have been told by scientists that humans are the wild card in future climate scenarios. What do they mean by that? Well, we understand the laws of science. When it comes to how Earth's systems will respond to increased greenhouse gases in the atmosphere and warmer land masses and oceans, there is the certainty of physics and chemistry. What scientists can't forecast is how humans will react, or what will happen to us. Will we continue combusting millions of years' worth of fossil fuels into the atmosphere? Will we rein in emissions? Will emissions only drop once large swaths of the planet become uninhabitable for humans and the population declines? At this point we have a great deal of control over what happens to the planet and what will be the future of our species and the many other species with whom we share the earth. The extent to which the planet will warm in the coming years and decades—and what impacts we'll experience as it does so—is our choice.

Right now Americans, and especially New Mexicans, are choosing to keep burning fossil fuels. In September 2019, according to the New Mexico Legislative Finance Committee (LFC), the state had 101 active drilling rigs, up from 72 the previous September. And production data from the Taxation and Revenue Department showed per-day oil production in the state was up 45 percent from the year before, while natural-gas production was up 21 percent from 2018. All that drilling means money for state coffers. Annually, the state receives billions of dollars in direct revenues from oil and gas production, which comes from severance and property taxes as well as royalty and rental income. Hundreds of millions more come from sales and income taxes on oil and gas drilling and services. The state also collects royalties from drilling on federal and state lands and imposes property taxes on the assessed value of products sold and on production equipment. In general, oil and gas revenue makes up 15 to 25 percent of the state's general fund revenue, depending on oil and natural-gas prices. And production is growing. And growing.

According to the US Energy Information Administration (EIA), oil production in New Mexico tripled in 2018 from 2009, with production exceeding 772,000 barrels per day. New Mexico became the fifth-largest oil-producing state in 2017, and it is also among the top natural gas–producing states. At the end of 2019 the state reported more than 125,000 wells in the state. This includes producing wells as well as those that had been approved.

These upward trends are predicted to continue. The EIA projected in 2018 that American oil production will increase to 10 million barrels per day in the early 2030s, up from 6.5 million barrels per day in 2018—more than 40 percent of which came out of the Permian Basin in eastern New Mexico and west Texas. Natural-gas production is rising in the Permian, too, due in part to higher prices, higher consumption, increased exports, and the construction of a new pipeline, the Gulf Coast Express Pipeline, in 2019.

During the summer of 2018 the US Bureau of Land Management (BLM) was readying to sell almost two hundred drilling leases for eighty-nine thousand acres in southeastern New Mexico, including about a dozen leases within a mile of the boundary of Carlsbad Caverns National Park. At that time Gary Kraft, a pilot who volunteers with EcoFlight, flew a handful of journalists (including me and my then twelve-year-old daughter), local ranchers, and activists above the Permian Basin.

From the air it was easy to see which lands are protected within the park boundary and which have been developed and industrialized. Thousands of feet below the Cessna, vegetation revealed the outline of the Black River. The crags of the Guadalupe Mountains seemed pressed almost flat in the June morning light. Sinkholes were visible too.

Even though it was early morning—and long before afternoon winds kick up dust—visibility from the airplane was poor. Looking down from the plane, the smoggy air reminded me of Los Angeles. Usually on trips like these I'm elated to be tucked into a small aircraft, gawking at the contours of the earth. This view, though, made me feel sweaty and nauseous. Each of the thousands of oil wells in the area has a road, and many have waste ponds. From above the surface of the earth, it was plain to see how hard this landscape is worked.

Since that time, even more wells have been drilled. Even more have been approved. And even more are in the planning stages.

Even the state's most progressive lawmakers and politicians—who acknowledge the impacts of climate change—haven't come to terms with how being an energy producer contributes to the greater problem, the global problem. As I write this in 2019, there are burgeoning state efforts to cut methane emissions from leaky or inefficient infrastructures in the energy industry. But with tens of thousands of existing oil and gas wells—not to mention pipelines, compressor stations, and tailpipes—and thousands more planned for the coming decades, it's not just methane we need to cut. It's also our dependence on fossil fuels more broadly. It's becoming increasingly clear that decisions around energy choices don't just require financial or political discussions. They demand moral conversations as well.

Glimpsing even a snapshot of what climate change means for New Mexico demands understanding the science of Earth's interconnected systems and diving into local, state, federal, and international politics. Reporting on climate change has meant walking across mountainsides burned by wildfires and hiking up dry riverbeds where desiccated fish flank the edges of the sandy channel. It's meant visiting with scientists and water managers, people who see oil-well flares from their front porches, and farmers who've wondered how they'll irrigate their fields next year as well as listening to faith leaders ponder moral responsibility. For me it has also meant having conversations about the "new normal" with my daughter and being honest with myself, and my readers, about what it feels like to report on an issue like climate change, which reaches into all of our lives, even when we try to ignore what's happening.

In this book I've condensed about fourteen years' worth of news stories and essays I reported while covering climate change in New Mexico between about 2005 and mid-2019. The introduction provides an overview of the impacts New Mexico and the southwestern United States are already experiencing and will continue to experience. Chapter 1 offers insight into what it feels like to confront the "new normal," and chapter 2 is a political overview that I based on my reporting during international meetings, three presidential administrations, and three gubernatorial administrations in New Mexico. Chapter 3 deals with energy development, specifically in the northwestern part of the state and on the eastern part of the Navajo Nation. Chapter 4 explores mourning forests and rivers and coming to terms with grief, and chapter 5 focuses on communities of faith and their efforts at climate-change activism. Chapters 6 and 7 are about heat and fire, including the devastating Las Conchas fire in the Jemez Mountains in 2011. Chapter 8 moves into the realm of water, specifically groundwater, while chapters 9 and 10 span the spring, summer, and fall of 2018 when the Middle Rio Grande dried in April and reservoirs were at record-low levels. By telling some of the stories of how warming already affects our rivers, forests, families, and futures, I hope we'll think more deeply about what's at stake if we lack the imagination to envision a future that isn't wrapped around the spokes of increased energy development, reliance on fossil fuels, and the embrace of economic growth at all costs.

Many segments of this book have previously appeared within a variety of outlets over the years, ranging from print magazines and online news outlets to public radio and *New Mexico In Focus*'s monthly program, "Our Land: New Mexico's Environmental Past, Present and Future." These various outlets include the *Santa Fe Reporter*, the scrappy alt-weekly that ran some of my earliest (and most favorite) pieces on climate change; KUNM-FM in Albuquerque; *New Mexico In Focus* at New Mexico PBS in Albuquerque; *New Mexico In Depth*; and—especially—*New Mexico Political Report*, where for two and a half years I had the good fortune to cover environment issues statewide, to my heart's content. All the sections previously published have been edited to make dates, definitions, or certain initiatives clear to today's readers.

While writing this manuscript and looking back on my coverage of water, climate, energy, and wildlife issues, it became so clear to me that while policies and politics changed, the most important issues did not. Human activities are causing the earth to warm. And if the scientific consensus has shifted at all over the past few decades, it has only been to solidify the link between humanity's greenhouse-gas emissions, the warming of the planet, and changes such as warming and rising oceans, melting sea ice and glaciers, changes in climate zones and ecosystems, and desertification. In other words, we know the climate is changing, we know it's due to human activities, and we know that humans and ecosystems will suffer the consequences. It's a bleak message, I know, for readers to hear when considering whether to keep turning these pages. But please know this: Along the way, I have learned more than I ever could have imagined about New Mexico and New Mexicans. And I've learned that New Mexico is worth fighting for, today and tomorrow.

It's my job as a journalist to talk with experts, question those in authority, and act as a proxy for the public. I have spent almost two decades doing that when it comes to environment issues in New Mexico, and if there is one lesson to be learned from each page of this book, it's this: We've known for a very long time what is happening and what is likely to happen, and we have done almost nothing to protect future generations from the impacts of climate change in New Mexico. If I weren't hopeful, however, that we could come together—stand back to back and face the challenges all around us—I wouldn't have bothered to share this book with you.

Acknowledgments

The first people to thank, always and in any situation, are my mother and my daughter. Though she hasn't always understood the paths I've chosen, my mom has never—not once—failed to offer unconditional support. She taught me about kindness and community and what it means to love ferociously and with a wide-open heart.

Then there is my daughter. From infancy through middle school, Lillie sat through day-long talks about water and climate change and endured marathon public meetings. I started bringing her along on reporting trips when she was only a few weeks old. Going back over files for this project, I found a recording from when she was just a few months old. With her strapped to my chest, we followed a scientist surveying for burrowing owls on a dusty afternoon; my recorder picked up her coos and gurgles that punctuated the interview with him. At age thirteen she even spent weeks without wireless access while I completely rewrote this manuscript in a cabin in northern New Mexico. She tolerated a nontraditional childhood without complaining—until she realized she could get a few laughs from her tales of hardship and woe. She still loves regaling people with the story of how she learned to make macaroni and cheese during the summer before first grade. (Working on deadline in my home office, I finally stopped saying, "In a minute . . ." to her requests for lunch when she popped in and asked, "Can you drain the noodles for me?") I love you, Lillie. Thank you for being my kid. You make every day an adventure in wild delight.

I'm grateful to the colleagues who edited, shot, or produced earlier versions of these stories, including Julie Ann Grimm,

Matthew Reichbach, Andy Lyman, Sarah Gustavus, Antony Lostetter, Kevin McDonald, Matt Grubs, Charlotte Jusinski, Jeff Proctor, Marjorie Childress, Trip Jennings, Elaine Baumgartel, Jodi Peterson, Cally Carswell, and Diane Sylvain. Thanks to friends at the Institute for Journalism and Natural Resources and to Frank and Maggie Allen. Also to the Earth Journalism Network and the European Journalism Centre. And thanks to everyone involved with running the Leopold Writing Program and maintaining Mi Casita, Aldo Leopold's cabin in Tres Piedras, New Mexico. Spending a quiet month at Mi Casita, away from the internet and with tens of thousands of acres of public lands out the back door, was truly a privilege. To say that month changed the course of my life is an understatement. And while the thoughtfully curated library full of books by and about Aldo Leopold helped inspire late-night thoughts and early-morning writing, I should be transparent and acknowledge that it was the ravens, Say's phoebes, western wood-pewees, turkey vultures, hummingbirds, common poorwills, Steller's jays, violet-green swallows, northern flickers, downy woodpeckers, and other avian hosts at the edge of the Carson National Forest who inspired me the most.

I'm grateful for the time each of my sources has spent with me. Any errors or misinterpretations are mine and mine alone. Special thanks is due to UNM's Dr. David Gutzler, whose rigorous academic work helps us understand what is happening to the earth's systems and our corner of the Southwest. Beyond his scholarly work at UNM and with the Intergovernmental Panel on Climate Change, Gutzler is a tireless advocate for his students and someone who never seems to turn down the invitation to speak to the public, to reporters, or to elected officials about climate change. Much of what New Mexicans understand about how climate change is affecting the state is thanks to his efforts at public education and outreach. This is the sort of work for which scholars and scientists are rarely compensated or praised—and his example is worth following. Thank you also to Collin Haffey, John Fleck,

Dave DuBois, Kerry Jones, Krista Bonfantine, Mary Carlson, Dagmar Llewellyn, Larry Rasmussen, Chris Hoagstrom, Thomas Archdeacon, Steve Davenport, Bradley Udall, Sen. Tom Udall, Jonathan Overpeck, Rep. Deb Haaland, and Albuquerque City Councilman Pat Davis, and to everyone willing to be interviewed at some point over the past decade. Thanks to Marcy Litvak, whose excellent Global Change Biology class in UNM's Department of Biology offers a rigorous introduction to atmospheric physics, Earth's carbon cycle, and plant biology. And thanks to the scientists and communicators at the National Center for Atmospheric Research, whose workshop on weather and climate modeling for journalists was useful and thought provoking. Thank you to the University of New Mexico Press's James Ayers, John Byram, and Elise McHugh. Also at UNM, I'm incredibly grateful to Dr. Maria Lane, Laurel Ladwig, and Hayley Pedrick for reading chapters, offering enthusiastic support, and supplying snacks along the way. I'm grateful, in fact, to everyone at UNM's Department of Geography and Environmental Studies. Academic departments can be notoriously competitive, soul crushing, and stifling. But thanks in no small part to Lane, GES is a place where people challenge and support one another and build relationships that matter, both professionally and personally.

I'd also like to acknowledge the public-information officers and attorneys working at state and federal agencies. Some of you were amazing, doing the job taxpayers believe you're doing: making sure that publicly funded research and policy is accessible to the press and to anyone else who wants to understand what's happening in their communities and in the world around them. Others, bowing to political pressure, personal feelings, or something else I don't understand, repeatedly acted against the interest of New Mexicans, refusing to answer phone calls, emails, or texts; delaying the release of information rather than facilitating it; and even removing information from government websites. I send a special kind of thanks to you folks: if you hadn't worked so hard to keep that information from me, I might not have realized just how valuable it was.

FIGURE 1. Jemez Mountains. Photograph by the author.

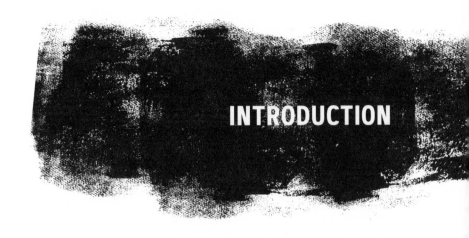

INTRODUCTION

CLIMATE CHANGE IS here. It's human caused. And it will deliver a blow to American prosperity. Already hard-hit by drought, wildfires, and declining water supplies, the southwestern United States will continue to face those challenges. And new ones.

That was the gist of a report released by the federal government in November 2018—on Black Friday, in fact. A day when Americans are supposed to shop, not read the news. The Fourth National Climate Assessment was part of the United States Global Change Program, a group first directed by Congress and then-president George H. W. Bush to report regularly on climate change. This particular assessment focused on impacts to America's economy and infrastructure.

Compiled by thirteen federal agencies and more than three hundred contributing authors, the peer-reviewed report reiterated what scientists had said for decades. It also clearly delineated the link between warming and extreme weather events and warned of the increasingly expensive economic consequences of not addressing climate change.

Economists project impacts in the hundreds of billions of dollars annually. And as the report's authors explained, "The assumption that current and future climate conditions will resemble the recent past is no longer valid."

In other words: where humans build their homes and highways, when we plant and harvest our crops, and how we manage forests or assess property values must all be considered through the lens of projected climate risk. We can no longer assume that systems, both climatic and human, will always work as they have in the past.

According to the assessment, rural and urban economies alike will suffer. Fisheries will decline, farming and ranching challenges will intensify, and rising sea levels will push cities and neighborhoods back from the coasts. American infrastructure, from highways and rail lines to sea walls and electric grids, will be harmed by impacts from climate change. Even trade, including import and export prices, will be disrupted.

The report also highlighted local or regional efforts to curb greenhouse-gas emissions and adapt to changes like rising sea levels and heat waves, and it offered resources and road maps for individuals, communities, and governments to boost these efforts. Adaptation, the authors wrote, means managing both short- and long-term risks. That's because even if humans immediately ceased greenhouse-gas emissions, the climate impacts due to past emissions will continue through midcentury. "Thus, climate risk management requires adaptation for the next several decades, independent of the extent of [greenhouse-gas] emission reductions," they wrote. "After 2050, the magnitude of changes, and thus the demands on adaptation, begins to depend strongly on the scale of [greenhouse-gas] emissions reduction today and over the coming decades."

Meanwhile, President Donald Trump tried to shore up support for his assertions—on the campaign trail and in office—that climate change is a "hoax." Two days before the federal report was released, the Twitter account of the president of the United States announced, "Brutal and Extended Cold Blast could shatter ALL RECORDS—Whatever happened to Global Warming?"

Compare that with what scientists are trying—again and again and again—to help the public and decision-makers understand.

When they released the assessment, the National Oceanic and Atmospheric Administration (NOAA) held a press call, summarizing the findings and giving reporters the chance to ask questions.

During that call David Easterling, director of the technical support unit of NOAA's National Centers for Environmental Information, offered what can only be described as a sobering overview of findings in the thousand-page report: "Observations of global average temperature provide clear and compelling evidence that global average temperature is much higher, and is rising more rapidly, than anything modern civilization has experienced," he told reporters. "This warming trend can only be explained by human activities, and especially by emissions of greenhouse gases into the atmosphere." Sea levels continue to rise, and extreme events—such as heavy precipitation—have increased and will continue to increase into the future.

These changes threaten the American economy, its infrastructure, the environment, and public health, he said—particularly in the absence of increased adaptation efforts.

"Climate change also threatens the many benefits the natural environment provides to society, such as safe and reliable water supplies, clean air, protection from flooding and erosion and the use of natural resources for economic, recreational and subsistence activities," Easterling said, adding that "future generations can expect to experience and interact with natural systems in ways that are very different than today."

Risks are highest for those who are already vulnerable, he explained, including low-income communities, some communities of color, children, and the elderly. "Future climate change is expected to further disrupt many areas of life, exacerbating existing challenges and revealing new risks to health—including mental health—and prosperity," he said. Easterling assured reporters that many impacts and damages could be "substantially reduced" through global-scale reductions in greenhouse-gas emissions complemented by local and regional adaptation efforts. Those efforts are underway, he said, but they don't approach the scale necessary to "prevent damages to the economy, the environment, and human health [that are] expected over the coming decades."

Some of the key points from the chapter on the southwestern

United States, which covers New Mexico, Arizona, California, Nevada, Colorado, and Utah, are worth highlighting:

- Temperatures increased from 1901 to 2016. Higher temperatures "amplify" recent droughts and contribute to "snow drought," which is when precipitation either doesn't fall or it falls as rain instead of snow. Continued warming will contribute to "aridification," a potentially permanent transition to an even drier environment than exists today. Depending on the amount of greenhouse gases humans continue to emit, the models forecast a number of different scenarios. Under the higher-emissions scenario, for example, the mountains in California currently dominated by snow could receive only rain by 2050.
- Human-caused climate change is contributing to water scarcity in the southwestern United States. Along with drought, demands from a growing population, deteriorating infrastructure, and dropping groundwater levels, warming is putting more stress on the region's already strapped water supplies.
- Forests and other ecosystems are having a harder time providing wildlife habitat, clean water, and jobs due to drought, wildfire, and climate change. The cumulative area burned by wildfire in the region increased between 1984 and 2015, and that burned footprint is twice what it would have been without rising temperatures. (The impact of rising temperatures on forests has also contributed to other factors including past forest-management practices.) With continued greenhouse-gas emissions, the Southwest will experience even more wildfires, the aftermath of which also contributes to flooding and erosion. In addition, warming is shifting where certain plant and animal species can live.
- The sea has already risen, and it has also warmed. Between 1895 and 2016, the sea level rose nine inches at the Golden Gate Bridge in San Francisco. Depending on emissions, by

2100 that change could range from 19 to 41 inches, and currently two hundred thousand Californians live within areas expected to be inundated. Warming in the Pacific Ocean also disrupts ecosystems, sickening or killing wildlife and harming commercial fisheries.

- Southwestern tribes face increased risk from drought, wildfire, and changing ocean conditions. Because the US government restricted some tribal nations in the region to the "driest portions of their traditional homelands," the well-being of southwestern tribes is at increased risk from water scarcity, the loss of traditional foods, wildfire, and changes in the ranges of plants and animals. Adaptation and mitigation measures are underway in many communities, but the authors note that "historical intergenerational trauma, extractive infrastructure, and socioeconomic and political pressures reduce their adaptive capacity to current and future climate change."

- Climate change affects energy demand and supply. As temperatures rise, demands for electricity rise. Yet demands for more water for fossil-fuel power plants will coincide with reduced water-supply availability from snowpack. Moreover, hydraulic fracturing in oil and gas drilling uses large amounts of water, pollutes water, and emits greenhouse gases.

- Climate change will increase future food insecurity. Southwestern farmers already worry about water shortages and grapple with increased drought and heat waves. As surface-water supplies become increasingly vulnerable, many areas have already depleted their groundwater supplies. Agricultural zones have already shifted, and this shifting will continue to affect where certain crops and fruit and nut trees can thrive. Together, these changes will cause geographic shifts in crop production, "potentially displacing existing growers and affecting rural communities."

- Southwesterners will suffer more health risks. The region

will continue heating, with more hot days and extreme heat events each year. People will also face more exposure to infectious diseases like the plague and hantavirus, experience more allergy problems, and be exposed to more severe dust storms.

The assessment authors also note that climate change affects mental health: "One impact of rising temperatures, especially in combination with environmental and socioeconomic stresses, is violence towards others and the self," the authors wrote. "Slow-moving disasters, such as drought, may affect mental health over many years. Studies of chronic stress indicate a potentially diminished ability to cope with subsequent exposures to stress." In the Southwest in particular, they wrote, the "loss of stability and certainty in natural systems may affect physical, mental and spiritual health of Indigenous peoples with close ties to the land."

Taken in total, the assessment for the southwestern United States offered very little good news.

"The big message for me is the interconnectedness between all these systems"—including water, food, energy, ecosystems, and human health—said Gregg Garfin, lead author of the chapter on the Southwest and a professor in climate, natural resources, and policy in the University of Arizona's School of Natural Resources and the Environment. "And if we try to look at those in isolation," he added, "we're probably setting ourselves up for more problems."

Here in New Mexico, many of our communities are already vulnerable. Even when other parts of the United States thrive, or recover from economic downturns, poverty keeps much of our state in its grip. Hand in hand with poverty we also have drug and alcohol abuse, political corruption, crime, homelessness, lack of access to medical and behavioral health care, child abuse, struggling schools, and deteriorating infrastructure and institutions.

Add to those long-entrenched problems the challenges that warming will bring: water scarcity, public-health problems, declines in agricultural yields, decreased recreational tourism, catastrophic wildfires, and, possibly, heightened animosity between urban and rural populations.

How well we adapt depends on how well we plan. And planning means taking care of one another.

———

THIRTEEN YEARS BEFORE the grim assessment came out, Garfin was among the scientists speaking at the third annual Drought Summit in Albuquerque. At the time, in the fall of 2005, he noted that the Southwest's projected population increases would also increase the region's vulnerability to climate change. At that meeting the University of Arizona's Kathy Jacobs urged managers and officials to connect long-term water planning to land use. She warned, "Don't set communities up for devastating failure." And the University of New Mexico's David Gutzler, a professor in the Earth and Planetary Sciences Department, was among the clearest voices warning of how warming would affect New Mexico.

Back in 2005 those climatologists, meteorologists, and biologists faced a near-empty auditorium. "There has to be a willingness to reconsider choices we have made," Garfin said to the audience at that time. "We must be able to cooperate, and be flexible beyond our wildest dreams."

Ten years after that, in 2015, the Indigenous People's Climate Change Working Group met at the Southwestern Indian Polytechnic Institute in Albuquerque. Within that group, American Indian student leaders work to ensure that Indigenous communities have a sustainable future as the climate changes and tribal communities lose their land bases, whether due to desertification or rising sea levels. They also work to retain access to traditional plants, historic fisheries, and ways of life. Already the homelands of Native people in Alaska and coastal Louisiana are washing away to rising seas.

One of the speakers at the meeting, Bob Gough, who is Rosebud Sioux, pointed out that humans are constantly adapting whether we realize it or not. Humans can adapt blindly, he said, or we can adapt to changing conditions consciously and deliberately with the understanding that what we build now—whether infrastructure, communities, or institutions—must operate in the future under conditions that will be different from what they are today.

What we do now will make a difference, he said. We shape the future with every decision and action we take. Gough paraphrased a quotation from Wayne Gretzky: "Don't skate to the puck. Skate to where the puck's going to be."

In New Mexico we know where that puck is headed. Especially when it comes to our rivers, streams, and aquifers.

In the arid Southwest, climate change "is all about water," explained Jonathan Overpeck, who has spent decades studying climate change and its impacts in the Southwest. Formerly at the University of Arizona, Overpeck is now the Samuel A. Graham Dean and Collegiate Professor at the University of Michigan's School for Environment and Sustainability. Despite the severity of his messages—he has hammered away at the science of climate change for decades—Overpeck is affable and approachable, willing to answer questions about science and opine about policy. He's also willing to start with the basics when it comes to why warming is a big deal for arid places like New Mexico.

Warming affects the amount of water flowing in streams, he explained, and the amount of water available to nourish forests, agricultural fields, and orchards. And a warmer atmosphere holds more moisture, demanding more from land surfaces. When it's warmer, plants need more water too. "Any way you look at it," Overpeck said, "water that normally would flow in the river or be in the soil ends up instead in the atmosphere."

To understand southwestern droughts of the distant past, scientists rely on tree rings as a proxy for rain gages and meteorological records. By studying tree rings—which offer a view of annual conditions and also summer versus winter precipitation—scientists have reconstructed rainfall records for the Southwest extending back more than 1,200 years. These climatic chronologies, based on tens of thousands of tree-ring samples, show that past southwestern droughts were notable for declines in precipitation, Overpeck explained.

Today's droughts are different, though. Even during wet years, which will still occur as the climate changes, warmer conditions dry out landscapes.

"With atmospheric warming, we're getting what we're calling 'hot droughts' or 'hotter droughts,'" he said. "That means that they're more and more influenced by these warm temperatures, and the warm temperatures tend to make the droughts more severe because they pull the moisture out of plants, they pull the moisture out of rivers and out of soil—and that moisture ends up in the atmosphere instead of where we normally like to have it."

Today's drought conditions, which Overpeck explained have been moving around the Southwest since the late 1990s, are exacerbated by warmer temperatures. When we spoke in 2018, the global temperature was 1.8°F higher than it had been in 1880. "What we're seeing now in the drought that's going on is that it's more due to temperature increase and less due to precipitation deficit," he said. And "hot drought" is what we should prepare to face in the future, too. "More and more so, the droughts will really be defined by hotness, by warm temperatures that just suck the moisture out of the soil, suck the moisture out of our rivers," he said. That, Overpeck explained, makes droughts even more devastating.

In 2017 Overpeck and Bradley Udall—cousin of US Sen. Tom Udall, a New Mexico Democrat—published a study showing that flows on the Colorado River between 2000 and 2014 averaged 19 percent below the 1906–1999 average. One-third of those losses were due to higher temperatures rather than changes in precipitation. They also wrote that if warming continues, the Colorado River will see 20 to 35 percent decreases in flows by 2050 and 30 to 55 percent decreases by 2100.

The next year, a follow-up study showed that even though annual precipitation increased slightly between 1916 and 2014, Colorado River flows declined by 16.5 percent during that same time period. That's thanks in large part to "unprecedented basin-wide warming."

Using experiments and a hydrology model, Udall, Mu Xiao, and Dennis Lettenmaier found that 53 percent of the decrease in runoff is attributable to warming; the rest is caused by reduced snowfall within regions that feed into the system. What's striking about the new study, Udall explained to me when it came out in 2018, is how much of the decline is due to warming relative to precipitation. "Climate change isn't in the future: it's here now, it's affecting all of us, and it will become increasingly worse as time goes on," said Udall, who is the senior water and climate research scientist at the Colorado Water Institute at Colorado State University. "Climate change is in our face right now: It's western fires, it's drought, it's river flows."

He cautioned that the study's results are based on one model and one data set. But, he said, the model is telling us something that's really valuable: "The feedback loop—self-reinforcing cycles, where dryness begets heat, which further begets dryness—is probably at the root cause of what's causing these 50 percent declines."

In short, he said, the model points toward the further aridification of the region.

The model also shows how sensitive the Colorado River Basin is to shifts in precipitation patterns. Not only does it matter whether precipitation falls as snow or rain, it matters where it falls. Snowfall in Colorado, Udall explained, contributes more to the river's flows than snow that falls in Utah. Unfortunately, the researchers noted a decline in snowfall within four important sub-basins of the river, all within the state of Colorado.

The paper, he said, is also a reminder that we need to respond to climate change right now. One of the biggest long-term issues is how to have an agricultural system that's resilient, even decades from now. "There's no way that the [sector] that uses 75 percent of the water cannot be in the crosshairs," he said. "We need to think about how to make that sector healthy, viable, and responsive to these coming changes."

He also cautioned that states and water users can't "engineer" their way out of the problems. In Colorado, for instance, there's talk of building more reservoirs at higher elevations, where there would be less evaporation. "It's a good discussion to have and a necessary one to have, but I'm very skeptical that engineering solutions are the ultimate answer," Udall said. "At some point, we're going to have to look at demand and how we manage to shed demand [in a way] that does the least amount of damage to communities and the environment and our economy."

But why does this matter to New Mexico?

The San Juan River, which flows through the northwestern corner of New Mexico, is a tributary of the Colorado River. Cities including Santa Fe and Albuquerque also rely on water piped from tributaries of the San Juan River through a series of tunnels and diversions and into the Chama River, which empties into the Rio Grande near Española in northern New Mexico. Farmers and homeowners within the Middle Rio Grande Conservancy District—from the Pueblo of Cochiti, through Albuquerque, and south on to the town of Socorro—rely on that imported water, too.

New Mexicans should also be worried about the Rio Grande, the state's largest river. "Its problems are pretty much the same as the Colorado River," said Overpeck. In fact, during the devastatingly dry year of 2018, water from the San Juan-Chama Project—and cooperation among the US Bureau of Reclamation, the water utility in Albuquerque, and local pueblos—was all that kept the Rio Grande flowing through Albuquerque that summer.

In all of this, there is good news, at least according to Overpeck. The good news, he likes to say, is that we understand what's happening and why it's happening. "We know that humans are causing the warming, and we know it's the burning of fossil fuels and the emissions of carbon dioxide that are the real culprit, so that's really important because it gives us a chance to stop it if we wish," he said. "That's the kind of debate we need to have in society: do we really want to risk losing half the flows in our rivers, or more? And if we don't, the good news is that we know how to slow down the losses just by slowing down our emissions of carbon dioxide."

Overpeck also likes to use a medical metaphor to lay out the "good news." If you know you're in bad shape, he asked, would you rather see a doctor who doesn't know what's wrong but sends you home with painkillers, or one who says, "You have cancer, that cancer is treatable, here's what we have to do to stop the cancer so you can live a normal life"?

"That's the way it is with climate change in the Southwest," he said. "We're going to lose water if we allow climate change to continue, but we know exactly why it's happening and what we can do to make it stop happening."

A few months after that conversation with Overpeck, in October 2018 the Intergovernmental Panel on Climate Change (IPCC) issued a report laying out just that—how to control future warming.

Ninety-one authors and review editors from forty countries prepared the special report, which noted that if humans don't drastically reduce greenhouse-gas emissions in the next decade, we won't stop warming, and that warming will have widespread and catastrophic impacts on Earth's ecosystems.

That's an urgent message from a body that works slowly and carefully and requires consensus.

The IPCC was created by the United Nations in 1988 so scientists worldwide could objectively assess climate-change data and help policymakers evaluate the state of the science of climate change. Since 1990 the IPCC has released a handful of assessments, on which thousands of scientists work together, reviewing and summarizing thousands of peer-reviewed studies related to various aspects of climate change.

For this special report, the IPCC was studying how a 2.0°C rise in global temperature will affect the planet, its ecosystems, and human communities, compared with a 1.5°C temperature increase.

Those numbers are important for one relatively arbitrary reason: Initially, countries of the world had discussed cutting emissions enough to limit Earth's warming to 1.5°C or less. Countries haven't made those cuts, however. Now the trick is to see how bad things will get if 2.0°C is the target.

What the IPCC found is that if Earth's temperature increases by more than 1.5 °C, the changes will be "long-lasting" and "irreversible."

The University of New Mexico's David Gutzler helped put the international report into perspective: "The more warming that we cause, the broader and more intense are the impacts in general," he said. "So, even for regions or components of the climate system for which the difference between 1.5° and 2° doesn't represent a big threshold or tipping point, the impacts of climate change get considerably bigger." Gutzler has spent decades studying climate change, in particular its impacts in the southwestern United States.

That's true here in New Mexico, where average annual temperatures have already increased by 2°F just since the 1970s. That faster rate of temperature increase here is due to the fact that continents warm more quickly than oceans, Gutzler explained. "The world is mostly ocean, and we are in the middle of a continent. So when people talk about global warming of one, two, or three degrees, for our region, we're thinking about double those numbers."

According to the IPCC's special report, limiting warming globally to 1.5 °C will require "rapid and far-reaching" changes in energy, transportation, agriculture, cities, land use, and industry. Globally, human-caused emissions of carbon dioxide "would need to fall by about 45 percent from 2010 levels by 2030, reaching 'net zero' around 2050," according to the authors, who also noted that "any remaining emissions would need to be balanced by removing CO_2 from the air."

The report was stark. And urgent. It grabbed headlines and caught people's attention. But its message wasn't altogether new. In the United States we have known about human-caused climate change and its impacts for decades.

For Gutzler, who is among the scientists working on the next IPCC assessment, the main message New Mexicans should glean from the special report and others like it is that climate change isn't coming—it's already here.

FIGURE 2. Lake Roberts, New Mexico. Photograph by the author.

1. THE NEW NORMAL

TUCKED IN THE far southwestern corner of New Mexico, the tiny town of Rodeo is a stone's throw from Arizona and just about twenty miles north of the US-Mexico border. It sits in classic Chihuahuan desert, with scented mesquite and creosote, and is framed by mountains—the Chiracahua Mountains rise toward the west, and the Peloncillos to the east.

It was the spring of 2011, and I had been traveling for work around New Mexico and to Colorado and the Dakotas. Everywhere, the symptoms of climate change were obvious, whether in the form of flood or drought. By early June I craved a few days to just enjoy a landscape without writing about it, and I sought an escape from Albuquerque. Even though it was technically still spring, it felt unbearably hot and dusty in the city. With an errand to run in Rodeo, my boyfriend and I were headed with my daughter for a few days into the Gila National Forest, a 3.3-million-acre stretch of conifers, trout streams, and dazzling canyons that encompasses the first wilderness area designated in the United States—in 1924, four decades before Congress passed the Wilderness Act.

When the waitress at the Rodeo Grocery & Café took our lunch order, the sky outside was blue. But in the time it took to power through a smothered burrito, the Chiracahuas to the west had disappeared. The wind had shifted, and smoke from the Horseshoe Two Fire in the nearby Coronado National Forest poured from the canyon, turning the sky a thick yellow and then scenting our two-hour drive north to the Gila with smoke.

It was dry in the Gila, too; runoff from snowmelt was about a quarter of the historic average that spring. But it felt good to wade in the East Fork of the Gila River, inspecting the undersides of river rocks for hellgrammites, tasting mint from an island in the river, and running our feet and fingers through the algae that streamed below the water's surface. (Hellgrammites, it must be noted, are awesome. They're larval dobson flies, and with their pinchy mouthparts, six legs, and additional pairs of filaments, they look like prehistoric monsters. They're terribly fun to look for in streams—and a sign of healthy aquatic habitat—though they'll also pinch the hell out of your fingers.)

Later in the day we drove past a burn scar from the Miller Fire that someone sparked at the end of April. Peering from the edge of the Heart Bar Wildlife Management Area, we saw how the fire scoured out the underbrush and streaked the trunks of conifers and cottonwoods black. The pines will be fine, but the cottonwoods growing at the banks of the river can't withstand the heat of fires like that. Those "old soldiers," as my boyfriend called them, would die.

By then it was all too clear to me: There was no running away from the issues of drought, fire, insect outbreaks, and climate change. Even the backcountry bears the signs of changes. Especially the backcountry.

The next evening smoke painted the setting sun a dramatic pinkish red over Lake Roberts, near the tiny town of Mimbres, where biologists stock the lake with trout ahead of the Aldo Leopold Kids Fishing Derby. Not long before sundown a game warden passed through with a young, cinnamon-colored black bear he had darted for relocation. With the forests crackling dry, bears were coming into towns or neighborhoods seeking food to sustain themselves. Asleep with a tag in his ear, the bear looked so small and soft that the kids wanted to stretch their hands into the cage and stroke his fur. But no one liked thinking about the future this slumbering bear, removed from its familiar ground, had ahead. Recidivism rates are high among hungry bears, and the forests were so, so dry.

The smoke cleared for Saturday morning's fishing derby, and droves of kids cast in the early-June heat. In the afternoon Lillie poked around the edge of the lake for frogs, delighted when two fisheries biologists pulled out their nets and helped her. Their friendliness and knowledge—and willingness to jump into the water after frogs—in tandem with her enthusiasm and sweet, high voice reset my heart and gave me the nudge I needed to return home to Albuquerque and get back to work.

Two days after we returned home, smoke from Arizona's Wallow Fire billowed just before rush hour across the West Mesa of Albuquerque and down into the city. The smoke lingered a week. We adjusted our daily schedules around winds and forecasts. The choke of the skies became routine.

Then, a few weeks later, one Sunday afternoon, I called Lillie to the north side of the house. Finally. Good signs! I pointed to a cloud that signaled the coming monsoon season. It was a far ways off, toward the Jemez Mountains. It certainly wouldn't bring rain to our neighborhood. But the enormous cloud over the mountains signaled the sweetest fruit of June in New Mexico—the storms that bring afternoon rains. Rains that release petrichor from our desert plants and make the air smell like sage and creosote. Rains that fill arroyos and nourish fields and gardens. Rains that tell us it's summer in New Mexico and reassure us they'll break up the heat at least a few times a week. We danced a booty-shaking, fist-pumping jig in the driveway and cheered the arrival of summer rains.

That night before bed, I checked the news.

About an hour from Albuquerque, in the Jemez Mountains, a fire was burning so wild and hot that it created its own weather above the mountains. When heat rises so fast that winds can't push it away, wildfires form a cloud—what's called a pyrocumulous cloud. From a distance these sometimes look like mushroom clouds. And they put firefighters—and anyone nearby—in danger. If the columns collapse, the falling air and smoke hit the ground and the wind spreads wildly, pushing the fire in new directions. Visible more than fifty miles away, that trickster cloud from the fire fooled me into thinking the summer's monsoon season had arrived.

The next morning I admitted my mistake and explained how very big fires—fires that grow almost forty-four thousand acres in a single day, like Las Conchas—can create clouds that resemble thunderheads.

This was the new normal, and from there on out I had to reevaluate what I thought I knew as fact before explaining our world to a five-year-old.

———

ELEVEN YEARS INTO a new millennium, New Mexico was warmer than it was even a decade ago. More than half of the state's native plants and animals have already been affected by climate change: some bird populations have shifted; some flowers bloom earlier. Sand dunes spread as the vegetation dies that used to anchor soils. Conifer forests suffer die-offs due to the continuing drought and booming population of bark beetles. Day to day, we have low river flows, dry soils, and wildfires. Worldwide, scientists watch their models and predictions play out—and we all experience symptoms no one expected. For example, no one expected the Arctic sea ice to melt as rapidly as it is.

As Las Conchas raged in the Jemez Mountains that spring and summer, Santa Feans had a front-row seat to a megafire that burned 156,000 acres. Patrick McCarthy, who long worked on climate-change issues in the southwestern United States, was shocked by the fire's ferocity. "No one has quite seen anything like this in New Mexico in terms of wildfire behavior, the speed at which it travelled, the incredible flame lengths," he said. "It was even creating its own weather in a roaring vortex."

McCarthy, the Nature Conservancy's director of conservation programs, also directed the Conservancy's Southwest Climate

Change Initiative. He had authored a number of reports detailing changes already occurring, predictions, and adaptive management strategies. He also worked on what used to be called "ecological management" and then, as everything was changing, started to be known as "transformation ecology," which represented a necessary shift of mind-set: Forests and other ecosystems can't be expected to return to what they were before something like a wildfire. The climatic conditions have changed, and therefore the same plant and animal species can't always be expected to return. As one of the first places hit by human-caused climate change in New Mexico, the Jemez Mountains have become a laboratory. "We're moving beyond denial to acceptance that we can't restore these forests anymore—because the forest of 50, 100 years ago wouldn't survive in 2050," McCarthy said in 2011. "So the question becomes: What is an ecosystem that's sustainable in 20 or 50 years?"

Watching Las Conchas rage in the Jemez Mountains was a difficult experience for many researchers, including McCarthy. "It was affecting places we care about, places we've been working on in some cases for decades," he said. Scientists are methodical, cautious. But bearing witness to the destruction of landscapes was an emotional experience for many of them, including McCarthy: "It's reached the point for me [where] the changes are so evident that they're really reaching deeply into my soul and how I feel about the relationship between humans and the climate, and especially our relationships with these places."

As certain ecosystems—like deep, dark pine forests—change, humans will need to decide where and how we live. "What kind of ecological processes that sustain people—like in Santa Fe or the Gunnison Basin, or Bear River in Utah—are we going to be able to sustain in the face of climate change?" McCarthy wondered. "You can walk out of the discussion, keep it at arm's length; but at strange, odd times, it comes back to me. It's poignant, riveting."

Climate change was already measurable, and it may also prove itself unpredictable.

"It's going to continue to bring this combination of projected effects and surprises that involve the crossing of these ecological or physiological thresholds that bring phenomena the likes of which we haven't seen before," he said. "There are things we can predict—droughts, larger wildfires—and there are things that are going to surprise us, things we can't predict."

It's clear that the changing planet calls for changes in human behavior. Unfortunately for us, humans are slow to change. Confronted with events that confuse or frighten us, we cling all the more fiercely to what we've known.

WHILE SANTA FE archaeologist Eric Blinman delivered a lecture titled "The Rear View Mirror: 2,000 Years of People and Climate Change in the Southwest," I sat rapt in the audience. New Mexico is an excellent place to study how people adapt to changes in their climate. Not only are there tree-ring and pollen records, but archaeological remains show where people lived and farmed.

Taken in tandem with climate data, Blinman explained, the archaeological record offers a comprehensive view of the past—of how in AD 300, when a prolonged drought began, it was too cold for farmers to move upslope to chase the rains. There is evidence of violence and of migrations to the south. Three hundred years later, a warming trend opened higher elevations to farming. As the moisture also increased, agricultural communities returned to the Four Corners. These sorts of shifts continue through time. They're plain to see in the record.

Climate change is both a crisis and an opportunity, Blinman pointed out during his lecture. He called the Galisteo Basin near Santa Fe the "poster child for climate change," then he began to trace its history. Beginning in the 1190s corn farmers, who relied on monsoon rains for their crops, moved into the Galisteo Basin. By the 1300s the hamlets had begun growing larger. A hundred years later the isolated hamlets had consolidated into eight large pueblos. This growth, Blinman said, was fueled by corn.

By around 1500 the system crashed. The climate pattern had returned to "normal," and significant rains were no longer falling in the Galisteo Basin. Unable to coax reliable crops from the high-desert basin, the communities could no longer survive in the area.

Blinman drew four lessons from these people who relied on corn and rain: Cultural expectations are abandoned with difficulty. People try to persist until too late. Social conflict and breakdown make the economy worse. And migration is the ultimate solution to climate change. I scribbled these down, copied them again onto a sticky note, and pressed the note to my computer monitor, where it hung, reinforced by tape, for years.

Just like the people who lived here centuries ago, we learn from the past. And we can also see what lies ahead. Today we have historical records showing that global temperatures have been rising. Measurement stations across the globe track carbon levels in the atmosphere and temperatures on land and at sea. Scientists have increasingly sophisticated models that show what's likely to happen as the planet continues to warm. Based on our understanding of how the earth and its atmosphere function, we know that the more carbon we add to the atmosphere, the warmer our planet will get. We lack good excuses to avoid planning. "The climate will change, regardless of cause," Blinman said. "And we have the potential for preventative adaptation."

———

EVEN FOR PEOPLE who understand that the connection between fossil fuels and climate change is not open to debate, it's convenient to imagine there is time left for a next generation to confront the issue. Despair and overwhelmed inaction are acceptable to some, as well. But all delusions and excuses disappear in the face of a man whose homeland is already disappearing into the rising sea.

In the fall of 2010 I attended a European Union meeting in

Brussels focused on climate change. European officials talked not just about their own emissions reductions and plans, but also how to get the United States to seriously and productively engage in international negotiations the following month in Mexico. There I met Eddie Osifelo, a journalist from the Solomon Islands, who said rising waters have already forced his people to migrate from the coast to the interior of the island. "Will you write about this?" he repeatedly asked me over the course of four days. "Will you let people know what is happening to us?" Fishermen learning to farm must also negotiate with the tribal leaders whose lands they now must share. That causes social conflict. It makes people fight with one another, he said. Distrust one another. And feel as though they don't have a proper place for themselves or their families anymore. The island and sea are changing, and so are people's identities and ties to their past, he explained to me. The people living on these islands did not cause climate change. They did not spend a century burning fossil fuels and building wealth. They did not call the seas to rise. And they don't know why the world doesn't help.

At a United Nations meeting a month after I met Osifelo, other men from the Pacific Islands talked about how their communities are losing ground to the ocean: "We will continue to be as noisy as we can until the water covers our heads," said Ronald Jumeau, the Republic of Seychelles's ambassador to the United Nations. "And even when the water is over our heads, when the bubbles pop, you will hear us yelling."

As global average temperatures have risen, African nations are experiencing ever-worsening droughts, food scarcity, and even flooding from extreme storm events. Asian countries that rely on glaciers to store and release water are worried about short-term flooding from accelerated melting in the Himalayas, but they are also concerned about long-term security threats because of water scarcity in the future.

In the spring of 2011 floods raged along the Mississippi and Missouri Rivers, and tornadoes shredded communities from Alabama to Missouri. At that point scientists would not attribute single weather events to global climate changes, but in mid-June, NOAA pointed out that only halfway through the year, the United States had already experienced eight disasters, each causing more than $1 billion in damages (for a total of $32 billion). That was before hurricane season had even begun—and it didn't include damages from drought. Drought was declared across the southern United States, from Virginia to California, with "exceptional" droughts occurring in Florida, across the Gulf Coast, up into northern Texas and Oklahoma, and on across southern New Mexico and Arizona.

On July 1, 2011, NOAA also released the "new normals." Based on data collected at more than 7,500 stations nationwide, normals are the thirty-year averages of climate variables such as temperature and precipitation. Meteorologists use them when forecasting weather; electric and gas utilities gauge short- and long-term energy-use projections. The 1981–2010 normals were about 0.5°F warmer than the previous thirty-year average.

The new normals are important, especially to farmers, who rely on consistent growing seasons, average temperatures, reliable spring runoff, and the start of monsoon season. On Ironwood Farm in the South Valley of Albuquerque, Chris Altenbach tries to adapt to extreme weather fluctuations. His livestock survived a deep freeze in early 2011. In Stanley, New Mexico, for instance, it went from 44°F on February 1 to 7°F the next day. By February 9 it was up to 41°F, and by February 20 the high temperature was 67°F. That cold spell was so low and mean that it "overwhelmed" electricity plants and natural gas–production facilities in Texas and New Mexico. People in northern New Mexico had their gas supplies cut when pipeline pressures dropped. And in southern New Mexico, companies implemented rolling blackouts to protect the overall systems.

Then the spring weather threw crops in New Mexico for a loop. The weather had been dry—the period from January 1 through the end of May was the driest on record at that time—but it was also unpredictable. On May 31, for instance, Altenbach and his neighbors awoke to a late freeze. By afternoon, temperatures soared to the 90s.

Variation like that squeezes crops and livestock, said Altenbach, who before taking a gander at farming worked as a biologist on the recovery of the endangered Rio Grande silvery minnow. It's tempting to say he has a soft spot for underdogs. And while the May freeze didn't kill his corn, it did slow its growth. The corn tasseled, he said, but the plants were only four to five feet tall—small for that time of year.

Adapting to the new normal, Altenbach gave up on his fruit trees—late freezes and drought make harvests unreliable—and decided to focus on annual vegetables. He still needs to build another greenhouse to grow more crops in controlled conditions, he said. He's also considering using row cover—hoops that hold fabric over the plants to moderate temperatures, increase humidity, and protect plants from insects—and he knows he needs to adjust infrastructure to deal with unanticipated freezes.

"I'm also doing more successional plantings," he said. "Rather than waiting for a certain time that I think is going to work, I'll plant a couple weeks apart just to try to make sure I get something to come in if there's an event that takes it out." Off the grid and committed to the practice of permaculture, Altenbach worried that, in having to adapt, larger operations will increase their dependence on fossil fuels. "By trying to mitigate for extremes in weather," he said, "other farmers are putting a little bit more strain on the environment." This reminded me of Blinman's four lessons and how past desert dwellers lost control of their societies.

Cultural expectations are abandoned with difficulty, people try to persist until too late, social conflict and breakdown make the economy worse, and migration is the ultimate solution to climate change.

When reporters wrap up interviews with a source, we reflect on the other pieces of the story: what other people have said and not said, what we've learned, and how all the stories fit together. There is no story bigger than the climatic changes the earth is grappling with right now. But it's also a story that people need to feel in order to understand. The story of our changing landscapes is written in the flames licking forests, in the topsoil that blows away when spring winds follow bone-dry winters. It's found within cornstalks and desert ruins.

And whether we like it or not, the story is also tied to politics.

FIGURE 3. Climate protest, Albuquerque. Photograph by the author.

2. THE OLD NORMAL, OR POLITICS AS USUAL

AS DONALD TRUMP delivered his inaugural speech on a rainy Friday morning toward the end of January 2017, I sat at my kitchen table in Albuquerque, streaming the inauguration and obsessively refreshing the White House website. Just a few minutes into the new president's speech, the website's transition appeared complete. All mentions of President Barack Obama's climate-change initiatives were gone. Over and over again I pasted the old whitehouse.gov links into my browser and clicked. The links were gone.

Gone. Gone.

Within hours of the inauguration the social media accounts for federal agencies sharply shifted messages. In 2016 many agencies, including the National Aeronautics and Space Administration (NASA), the US Department of the Interior, and the US Geological Survey (USGS), seemed to have ramped up posts related to climate change. On a regular basis federal employees were posting about things like the extent of the Arctic sea ice, drought, worldwide temperatures, carbon levels, and newly published studies related to climate. In the Trump administration that outreach to the public on climate, especially from high-profile agencies like NASA and the National Park Service (NPS), slowed or stopped.

While state and local governments can enact and enforce laws and rules related to climate change, federal policy—on everything from renewable energy to clean water—drives both innovation and compliance. New Mexico is particularly vulnerable to federal policy due to a number of factors, including the state's huge federal workforce and its millions of acres of public lands.

Within days of the inauguration, the White House announced a federal hiring freeze. Trump's cabinet appointments included the CEO of ExxonMobil, Rex Tillerson, for Secretary of State, and the new administrator of the US Environmental Protection Agency (EPA) was an Oklahoma state attorney general, Scott Pruitt, who had repeatedly sued the agency.

Immediately, the Trump administration set about rolling back the few gains made late in the Obama administration. As carbon dioxide hit levels unseen in 650,000 years and global temperatures continued climbing, Trump signed an executive order revoking and rescinding Obama-era orders and reports addressing climate change and clean energy. That order, on "promoting energy independence and economic growth," was just one aimed specifically at boosting oil and gas drilling and lifting "regulatory burdens" on industry. He ordered the EPA to review and revoke the Clean Power Plan, which required states to cut greenhouse-gas emissions from power plants. Trump also rescinded guidance requiring federal agencies to consider the impacts of greenhouse-gas emissions when performing environmental reviews, disbanded the Interagency Working Group on Social Costs of Greenhouse Gases, and withdrew the group's documents "as no longer representative of governmental policy."

Trump directed Interior Secretary Ryan Zinke to review his agency's rules, including one guiding hydraulic fracturing on federal and Indian lands. Zinke followed up with more orders of his own. One overturned a moratorium on new coal leases on federal lands, another directed an agency-wide review of all existing regulations, documents, and policies that might hinder energy development, and a third created a new committee to advise Zinke on royalties collected from mining and drilling on public lands and in US waters. And, of course, as many New Mexicans know, at the behest of Utah

Sen. Orrin Hatch, Trump ordered Zinke to review the boundaries of millions of acres of national monuments including two in New Mexico. (The recently designated Organ Mountains–Desert Peaks National Monument and Rio Grande del Norte National Monument avoided the chopping block, but the administration shrunk the boundaries of other monuments, such as Bears Ears and Grand Staircase–Escalante to make way for resource extraction.)

That all happened within one hundred days of Trump's inauguration.

There's no doubt that the Trump administration's actions are extreme. But this wasn't the first time science was under siege.

During the administration of George W. Bush, Vice President Richard Cheney and Deputy Chief of Staff Karl Rove centralized White House control over agencies in a way that hadn't happened before. The administration became notorious for rewriting scientific documents and holding frequent briefings and pressuring high-level agency employees. As a reporter for *High Country News* at the time, I watched that play out in New Mexico. In 2001 Fish and Wildlife Service biologists trying to protect endangered silvery minnows in the Rio Grande had issued a "biological opinion" requiring water managers to keep water in the river for the fish, even if that meant cutting irrigation water to farmers. Pressured by political appointees, the next year Fish and Wildlife rescinded that opinion and then issued a new one, with altered flow recommendations. Biologists were "under fire"—marginalized or else expected to change their recommendations.

The Bush administration exerted such extreme control over issues related to science and the environment that early in Obama's administration he issued a memo on scientific integrity to the heads of executive agencies and departments. Obama directed each agency to draft scientific integrity policies that would strengthen the credibility of government research. The policies, which included barring public-affairs officers from directing scientists to alter their findings, were meant to boost public trust in science and the scientific processes informing policy decisions.

However tempting it is to look back to the Obama administration with a sense of wistfulness, it's important to be clear-eyed about the fact that Obama's administration rarely championed environmental issues or the agencies and employees working on natural resources. Energy development on public lands boomed under Obama, and environmental enforcement continued its decline during his administration. The lead and water crisis in Flint, Michigan, underscored a lack of federal attention to local drinking-water issues and, for the most part, biologists with the US Fish and Wildlife Service continued focusing on reviewing new development projects rather than doing what could be considered "deep science." And it was during the Obama administration that the Fish and Wildlife Service did away with flow requirements in the Rio Grande for endangered species in New Mexico.

It's also important to note that the US government had known for decades that burning fossil fuels was having a harmful effect on Earth's atmosphere.

More than a half-century ago, scientists explained to President Lyndon B. Johnson that the burning of fossil fuels was increasing carbon-dioxide levels in the atmosphere. In 1965 Johnson's science advisory committee wrote in a White House report, "Through his worldwide industrial civilization, Man is unwittingly conducting a vast geophysical experiment. Within a few generations he is burning the fossil fuels that slowly accumulated in the earth over the past 500 million years." The carbon dioxide humans were injecting into the atmosphere would cause changes, they wrote, that could be "deleterious from the point of view of human beings."

———

IN FEBRUARY 1965 Johnson gave a special message to Congress, focused on conservation. "For centuries Americans have drawn strength and inspiration from the beauty of our country," he began. "It would be a neglectful generation indeed, indifferent alike to the judgment of history and the command of principle, which failed to preserve and extend such a heritage for its descendants."

The president also spoke of rising carbon-dioxide emissions: "This generation has altered the composition of the atmosphere on a global scale through radioactive materials and a steady increase in carbon dioxide from the burning of fossil fuels," Johnson said.

And yet the United States has spent decades, even during the Obama administration, thwarting meaningful international action on climate change. Our refusal to cut emissions or come to the international table with anything resembling humility or cooperation has even influenced other countries like Canada and Russia to back away from emissions-reductions commitments they made under the Kyoto Protocol in the 1990s. In 2010—during the Obama administration—I spent two weeks reporting on the annual Conference of the Parties of the United Nations Framework Convention on Climate Change in Cancún, Mexico. (The United Nations has convened annual meetings on climate change since 1995.) And I watched Todd Stern, the US special envoy for climate change, spin politics in daily press briefings while scientists warned of what was happening to the planet and people from island nations begged the world for help.

During opening ceremonies, the IPCC's chairman, Rajendra Pachauri, asked negotiators to heed the science. "The warming of the climate system is unequivocal," he said. As many as 30 percent of the plant and animal species assessed so far are at risk of extinction if the planet exceeds the 1.5 to 2.5°C rise in temperatures, he said, adding that global carbon emissions should peak no later than 2015 and decline thereafter. If humanity seeks a chance at averting abrupt and irreversible climate change, those changes needed to occur, he said in 2010. Already some impacts are inevitable, he said, such as sea-level rises from the melting of sea ice that has already taken place.

Pachauri went on to talk about adaptation and mitigation,

pointing out that the response to climate change demands an integrated risk-management process—and that changes of lifestyle and patterns can contribute to climate-change mitigation. For a moment during an opening ceremony, Pachauri pleaded with negotiators, asking them to take significant action while in Cancún: "The available scientific knowledge justifies it," he said, "and the global community rightly expects it."

But throughout the two-week meetings, politics' triumph over science was evident. And the United States led the charge.

After eight years of the George W. Bush administration's inaction, nations worldwide embraced Obama's election and a Democrat-controlled Congress as a sign the United States would finally step up on climate. In December 2009 more than 190 countries signed the so-called Copenhagen Accord, agreeing that worldwide temperature increases should not exceed 2°C. But the accord—negotiated by heads of state and not under the rules of the convention—did not actually commit to achieving that goal by cutting carbon emissions. It failed to yield any legally binding agreement on cuts. And the United States did not pledge any substantive action.

That was what the meetings in Cancún were supposed to build upon.

Meanwhile, Stern reiterated the same message in his press briefings—"The US seeks a balanced package of decisions"—and deflected the rare, pointed question from the press. After two weeks the meetings ended without agreements or commitments. Delegates from every country except Bolivia agreed again to delay action on everything from carbon-emissions reductions to funding adaptation projects in developing countries, where the impacts of climate change are most severe.

That morning I shared an airport shuttle with a woman from South Africa. Throughout the two weeks I'd met people from around the world who were suffering the impacts of climate change—watching their islands diminish beneath rising seas or worrying about melting glaciers in the Himalayas. Looking at my press badge, they'd ask me to please write stories that will show Americans what is happening to them. Time and again, whether riding the bus, standing in security lines, or eating dinner, people from across the world asked what was happening in America. They didn't understand why the US government for decades challenged a binding international agreement to cut carbon emissions, mitigate the impacts of climate change, and help developing nations deal with drought, rising seas, or floods that destroy infrastructure. People from Senegal, Ghana, Zimbabwe, Panama, Colombia, India, and across the European Union asked me why the American people don't care about climate change.

The South African woman in the shuttle was disappointed by the negotiations—"They agreed last night that they were pleased to be going home," she said—but she was not surprised by the Americans. Adapting to climate change isn't about saving the environment, she said to me. It's about economics and survival. But the American people tell their leaders not to act. In a democracy, after all, elected leaders carry out the wishes of the citizens.

Despite our notions of American exceptionalism, most of the people I met in Cancún in 2010 knew the US government would never lead on climate change. They were looking to other countries, including China. But people were still curious why Americans fall for the histrionics of industry-funded climate-change deniers when it's clear what's happening to the planet. They wondered when Americans would pay attention to the rising seas, the Amazon's burning forests, and the glaciers melting from mountains across the globe. They wondered why Americans don't demand their government become a responsible member of the international community.

They were wondering. But they're not waiting. That became painfully clear to me as I headed toward the airport with my fellow South African passenger. "Everyone used to want to be like America," she said. "I think that is not the case anymore."

Since then, of course, things have gone further downhill. In June 2017 Trump announced that "in order to fulfill my solemn duty to protect America and its citizens, the United States will withdraw from the Paris Climate Accord." That accord was signed in December 2015, when countries agreed to reduce greenhouse-gas emissions worldwide. The deal lacked enforcement mechanisms, and under it pledges to cut carbon emissions are voluntary. But it was a clear signal that climate change required immediate, and concerted, attention. At that meeting countries also agreed that previous plans to limit warming to 3.6°F (2°C) would not protect many countries, especially island and coastal regions.

While the agreement was certainly flawed—too little, too late, and with no real enforcement—the Trump administration's rhetoric about the agreement and how it would affect Americans is notable for its obfuscation.

Trump criticized the agreement for putting US workers at a disadvantage to other countries and leaving American workers—"who I love," he said—and taxpayers to absorb the costs. "As of today, the United States will cease all implementation of the non-binding Paris accord and the draconian economic and financial burdens the agreement imposes on our country," Trump said to a cheering crowd. He added that the United States would no longer contribute toward the Green Climate Fund, which he said "is costing the US a vast fortune." The fund finances renewable energy and infrastructure in poor countries that are particularly vulnerable to the impacts of climate change—impacts caused by the greenhouse-gas emissions of industrialized countries like the United States—and it was created by 194 countries in 2010. Since the nineteenth century, it's worth noting, the United States has emitted more carbon dioxide than any other country on Earth—more than 28 European countries, including the United Kingdom, combined.

Trump added that the Paris Accord failed to live up to his administration's environmental ideals. "As someone who cares deeply about the environment, which I do, I cannot in good conscience support a deal that punishes the United States, the world's leader in environmental protection, while imposing no meaningful obligations on the world's leading polluters," he said, calling out specifically India and China. The bottom line, he said, was that the Paris Accord was "very unfair, at the highest level, to the United States."

Just minutes after the Trump administration announced its plans to withdraw from the agreement, I met with New Mexico Sen. Tom Udall at his office in downtown Albuquerque. A long-time champion of environmental issues—as a senator, a congressman, and, before that, as New Mexico's attorney general—Udall is also the son of Stewart Udall, who served as Secretary of the US Department of the Interior under presidents John F. Kennedy and Lyndon B. Johnson. The elder Udall wrote *The Quiet Crisis*, about the need for a national "land conscience," in 1963; two years before he died in 2010, he wrote a "message to our grandchildren." In that open letter he noted that their generation will "face a series of environmental challenges that will dwarf anything any previous generation has confronted," and he called for energy efficiency. He also wrote of the "moral responsibility" of the United States and China to cut carbon emissions.

New Mexico's long-standing senator, Tom Udall was one of ten US senators who attended the United Nations climate talks in Paris in 2015, and he repeatedly urged Trump to remain a party to the agreement. Abandoning the agreement, he told me, was "one of the worst decisions" the United States had made in a long time. "It could be the worst," Udall said. He acknowledged criticisms of the agreement, but he said it represented all countries trying to make a contribution toward stemming climate change. Backing away hurts the stature of the United States, he said, adding that, "I don't think the world is going to stop for us."

"I think the reaction—and there may be many unintended consequences—is to punish us in some way. To say, 'Well, if you're not going to work with us on this, we're not going to work with you on that,'" he said. "We don't know how all those things play out, but I think it hurts us economically, it hurts us on all the trade issues out there, the good relationship we have."

Not taking action on climate change also harms New Mexico,

one of the places in the United States where the impacts of rising temperatures—due to rising greenhouse-gas emissions—is most pronounced. The Southwest is among the regions in the United States that are most vulnerable to climate change. Already, warming is affecting surface and underground water supplies, increasing large wildfires, and causing large-scale tree die-offs along with other environmental, economic, and public-health impacts. New Mexico is currently the sixth-fastest-warming state in the United States, with average annual temperatures expected to rise 3.5 to 8.5°F by the year 2100, according to a 2016 report released by the Union of Concerned Scientists.

In contrast to the president's assertions that sticking with the Paris agreement would harm the US economy, Udall said the United States is "missing the biggest business opportunity there is for the next century and into the future." Other countries are going "full bore," he said, and committing to renewable energy and new technologies. Venture capitalists have long asked the US government for three things, he said: an imposed price on carbon, a nationwide renewable energy standard, and a tax code that has long-term incentives for renewable energy.

"Then people would see, we are really committed to this. The United States of America is committed, we're all in," he said. Then private investors would commit money to research and development—and make a difference. "That's what we're missing out on now," he said. "It doesn't mean that a researcher or somebody who has a little bit of money might not continue to do it, it's just that we had the potential to lead the world. That's what we're giving up: the potential to lead the world."

To be sure, the United States could have chosen to lead a long time ago.

"My dad was right in the middle of that," Udall said of the Johnson administration's realization of how industrial development was affecting the planet's atmosphere and climate. His father, Stewart Udall, was first elected to Congress in 1954, representing Arizona. He also led the US Department of the Interior from 1961 to 1969, under President John F. Kennedy and, later, Johnson. Udall was an advocate for expanding public lands, including national parks and monuments, and he was an integral player in laws like the Wilderness Act of 1964, the Endangered Species Act, the Clean Air and Water Acts, and the Wild and Scenic Rivers Act.

By the time Stewart Udall was serving as the interior secretary, the American public had awakened to environmental issues, said Udall, and the passage of early environmental laws was bipartisan. "That was the thing I think my father was saddest about as he grew older," Udall said. "He died in 2010, and I'd say his last ten years—from 2000 to 2010—he just couldn't believe how the environment had become partisan."

The turning point, Udall said, may have been what's called the Powell Manifesto. In 1971, just before he was nominated to the US Supreme Court by President Richard Nixon, Lewis Powell wrote a memo to the head of the US Chamber of Commerce.

Powell wrote of the "assault on the enterprise system" by "Communists, New Leftists and other revolutionaries." Businesses, he wrote, had only responded by "appeasement, ineptitude and ignoring the problem." He proceeded to lay out what individual companies and the Chamber of Commerce needed to do, including through public-relations departments, by demanding equal time in the media and on college campuses and by "evaluating" textbooks.

Powell also wrote of the "neglected" political arena and courts. "As unwelcome as it may be to the Chamber, it should consider assuming a broader and more vigorous role in the political arena," he wrote. He added that the chamber should follow the example of civil-rights and labor groups. "Their success, often at business' expense, has not been inconsequential," he wrote, adding that the courts provided a "vast area of opportunity" for the Chamber of Commerce.

Overall, business needed to become more aggressive, he wrote. "It is time for American business—which has demonstrated the greatest capacity in all history to produce and to influence consumer decisions, to apply their great talents vigorously to the preservation of the system itself."

Udall sums up Powell's memo more succinctly. "He said, 'You guys are getting killed. You should all organize, and you should get into Washington and you should use all of your power and your might to fight the things that are coming out of Washington,'" Udall said. "They were talking about a lot of the regulatory issues, but they were also talking about the conservation and environmental laws that had been put into place."

Presidents and Congress shied away from, or resisted, action on climate change. And environmental and conservation issues became increasingly political.

In the 1970s Congress set limits on campaign spending, some of which were struck down by the US Supreme Court, and also increased authorization of political action committees, or PACS. Then in 2010 the Supreme Court upheld a case arguing that the government can't restrict free speech by limiting campaign contributions from corporations, nonprofits, and labor unions.

Now, Udall said, spending is out of control. And corporations are still helping squelch regulations, environmental laws, and movement on climate change. "You hear everybody now talking about how the system is 'rigged.' Well, it is. They've tipped it in favor of the corporations and the wealthy," Udall said. "And they're impacting government."

By ignoring the warning signs of climate change, Udall said that Trump and congressional Republicans are putting the environment, the economy, and children's futures at risk. "Families across New Mexico see the impacts of global warming every day in the form of rising temperatures and extreme weather. Last year [2016] was the hottest year on Earth—and the third consecutive year to break global temperature records," he said, adding, "President Trump called the Paris Agreement a 'bad deal'—but the real bad deal is saddling our kids with more drought, wildfires and rising oceans and temperatures."

AND THEN THERE'S the political situation in New Mexico. The same year of the Cancún talks, New Mexicans elected Gov. Susana Martinez, a Republican, who sailed into office with strong support from the oil and gas industry.

Her predecessor, Democratic governor Bill Richardson, had emphasized renewable energy, directed state agencies to study the impacts of climate change, and was inching toward policies that would protect groundwater and public health from the impacts of oil and gas drilling and coal-fired power plants. Richardson, who served from 2003 to 2011, signed a statewide Renewable Energy Portfolio into law in 2004 and issued executive orders calling for climate studies and the reduction of greenhouse-gas emissions. In 2007 he and four other western governors, including California governor Arnold Schwarzenegger, a Republican, sought to develop a market-based program that would reduce greenhouse-gas emissions. Early in Richardson's administration the state released its study about the potential impacts of warming, including increased public-health problems, water scarcity, wildfires, and infrastructure damage from severe storms as well as a decline in tourism. The forty-seven-page report provided a road map for state leaders to confront the challenges warming posed to New Mexicans.

During that time, state legislators also started taking the issues seriously. In 2007 alone they introduced twenty-four bills and memorials related to climate change. In comparison only twenty-two bills even mention the phrase between 2011 and 2017.

Immediately upon taking office Martinez fulfilled her campaign pledges to industry and set about revamping state agencies, rescinding rules related to water protections, and dropping discussion of climate change from all state programs and departments. She withdrew the state from the Western Climate Initiative, ended efforts by the state's Environmental Improvement Board to cut carbon emissions, revamped and weakened the New Mexico Environment Department, directed the elimination of climate-related programs, and issued an executive order creating a task force to "identify all red tape regulations that are harmful to business growth and job creation."

Her initial choice to lead the state's Energy, Minerals and Natural Resources Department was former US senator Harrison Schmitt, a climate-change denier who eventually withdrew his name from consideration to avoid a background check. And the secretary leading that department during her final years in office was the retired vice president of WPX Energy, a top energy producer in northwestern New Mexico. One of her picks to lead the New Mexico Environment Department came from a law firm known for its representation of mining companies; when he left the state job, he went on to direct the New Mexico Oil and Gas Association.

Even as firefighters tried to hold back New Mexico's largest wildfires in the state's recorded history, as rural communities were whacked by drought, and as the state's largest rivers dried, Martinez ensured that agencies rolled back their climate initiatives—even canceling a Climate Masters Class for the public—all the while avoiding the words "climate change."

And yet the signs of climate change were everywhere around the state. Rural communities were being pummeled by drought. The Rio Grande and the Pecos Rivers dried, and everyone from farmers to rafters worried about snowpack and river flows. Tribal communities were disproportionately affected by both drought and post-fire flooding—and they weren't receiving federal emergency funds due to hold-ups at the state agency responsible for passing them through from the Federal Emergency Management Agency. And New Mexico experienced a number of historic fires, including Las Conchas in northern New Mexico in 2011. In 2012 in the Lincoln National Forest, the Little Bear Fire burned 44,330 acres—and 254 buildings. In southwestern New Mexico, the 2012 Whitewater-Baldy Fire burned a whopping 297,845 acres. And in 2013 the Silver Fire burned another 138,000 acres in and around the Gila National Forest.

In June 2017 Martinez joined incident commanders and US Forest Service officials to update local residents on yet another forest fire. She had already declared a state of emergency and called in the state National Guard for the Dog Head Fire in the Manzano Mountains near Albuquerque.

Since taking office five years earlier, Martinez had presided over similar meetings across New Mexico while various fires burned millions of acres of forest. With warming, the number and size of fires has increased, and fire season had lengthened by about two months in the western United States. Even a 2005 report prepared by the state warned that New Mexico's forests were vulnerable to "catastrophic" wildfires and massive diebacks.

But when asked at that Dog Head Fire meeting about her plans for long-term forest management, Martinez complained about the federal government. "There are very large amounts of land that belong to the US Forest Service, we can't as a state control it," she said to me when I waited in line following the public meeting.

Martinez refused during her entire administration to answer my questions; for years her communications and legal staff even refused to acknowledge voicemails and emails requesting interviews, comments, and information. In eight years of her administration, the only time I was ever allowed to speak directly to her was during this public meeting, when I stood in line with people eager to meet the governor or ask a question about emergency services. "They're not maintaining them," Martinez said to me of the national forests. "They maintain them for this critter or that, the spotted owl, or the sand dune lizard, whatever it may be that they're listing as an endangered species, and so you don't treat the forest and clean it out and thin it out. There is an enormity of fuel."

The state has "argued" and "fought" with the Forest Service, she said. "So, it's like, you know what? Give it back to us and let us take care of it," she said. "Then we will be able to take care of it much better because it affects us."

For the course of her entire administration, Martinez and her appointees dodged and ignored questions about her position on climate change and the administration's plans to prepare the state for warming's impacts on water supplies, agricultural yields, forests, and public health. The governor, whose election and reelection campaigns were heavily financed by companies and individuals in the oil and gas industry, didn't publicly deny climate change or call it a "hoax" like some other Republicans, including Trump. Nevertheless she effectively thwarted the state's ability to address the impacts of climate change and take advantage of economic opportunities in renewable energy.

It was during her administration that New Mexico came under international scrutiny in 2014, when scientists published a paper showing that the largest methane anomaly in the United States hovers over northwestern New Mexico. The satellite image of the methane plume splashed across the national and international news, and it's easy to see why: in the satellite images, the Four Corners showed up as an ugly welt of yellow, red, and orange surrounded by cool greens and blues. Colorado and New Mexico convened public meetings, and a number of the study's scientists, including Eric Kort of the University of Michigan and Christian Frankenberg with NASA's Jet Propulsion Laboratory, came to Farmington to answer questions.

Eric Kort, the lead author on the methane paper, urged caution when interpreting the image. "It's easy to look at that image and think that it means that the most emissions are coming from this region," he said. "But it's not an image of emissions, it's an image of concentrations of the gas. It does not at all mean that it's the highest emissions region. Those are different things."

What they see over the Four Corners is more methane than in the surrounding area, he explained. That's due to "both the emissions in the region and the nature of the winds, so those emissions tend to stay concentrated there a little more."

Scientists from NOAA also worked to figure out where exactly the methane was coming from. Owen Sherwood, with the Institute of Arctic and Alpine Research at the University of Colorado–Boulder, was part of that team, and he spent time driving around the highways and back roads of the Four Corners. His truck was outfitted with special equipment and a long mast on top. The whole thing was engineered by the lab manager where Sherwood works. "For better or worse," laughed Sherwood, lab manager Bruce Vaughn has "leant his truck to the services of doing science."

Different sources of methane have different signatures, Sherwood said. If you study those, you can figure out whether the methane is from a feedlot or from oil and gas drilling. Or even from forest fires. "They all have unique fingerprints," he said. "And we can use that information to discern where the methane is coming from." While Sherwood collected samples on the ground, Christian Frankenberg, with NASA's Jet Propulsion Laboratory and another author of the methane-anomaly study, took to the skies, collecting samples by flying through the plume.

By 2016 they had some answers: they identified 250 emitters of methane—which is the same as the natural gas people use to heat their homes and cook their dinners. The sources included leaky pipelines and storage areas, natural vents in coal beds, and a new natural gas–processing facility—and they found that 10 percent of those 250 were responsible for about half the methane emissions in the San Juan Basin.

Methane is a greenhouse gas, and it is thirty to eighty times more potent than carbon dioxide. That means all this development affects the climate. But the development also affects people who live in the area. The new wells light up the night skies with flares. Trucks rut the roads, drive up traffic, and create dangerous conditions. There are pollution and safety hazards. But the development also causes conflicts within communities. It changes how people relate to one another, and even how they relate to their ancestors and envision their futures.

In 2015 the nonprofit group Diné Citizens Against Ruining Our Environment, or Diné CARE, was among the groups suing the Bureau of Land Management and the US Department of the Interior for the federal government's approval of 130 new oil wells in

New Mexico near the eastern edge of the Navajo Nation. Lori Goodman, along with a number of Navajo women including Etta Arviso, fought against the new wells, against the encroachment of drilling upon Chaco Culture National Historical Park, and against a new pipeline planned to carry fifty thousand barrels of oil a day from the Four Corners south to along Interstate 40.

It wasn't an entirely new battle for women like Goodman, who have spent decades fighting coal companies and other industries from exploiting the Navajo Nation—its people, lands, and waters. Spread across more than twenty-seven thousand acres in New Mexico, Arizona, and Utah, the Navajo Nation has long been targeted by industry and the US government for its natural resources, including coal, uranium, and natural gas. Even now, Goodman explained, many Diné people living on the reservation don't have access to clean, running water—they haul it in tanks. And yet the tribal government supplied low-cost water to local coal mines and coal-fired power plants. Tens of thousands of acre-feet of water each year go toward the extractive industries and for power-plant cooling towers, she said. The tribal leadership doesn't understand the value of water, she said. They've signed away water rights and sold water to industry instead of planning for the future: "The water is always for extraction; they're using it for fracking, for power plants, for mining. Those are all the water uses, and we're supposed to be getting rich from them," she said. "But it hasn't happened in forty years, and it's not going to happen."

Far-reaching decisions are made on international and national levels. And royalties and revenues from oil and gas drilling make up a significant portion of New Mexico's state budget. But the local communities and families bear the direct burdens associated with development, whether that's traffic and noise or public health or environmental impacts. "It's time for people in Albuquerque [to realize], your children are impacted, it's your future, too," Goodman said, walking past piñon trees and sandstone outcrops and heading down from the top of a hill a few miles south of Highway 550, where at least four drilling rigs mar the sweeping view. "You can't just think, 'Oh, it's just happening to some isolated group of people.' No, we all breathe the same air, and we all need to stand up and protect our environment. There are better ways of doing this. And this is not the right way."

FIGURE 4. Counselor Chapter, Navajo Nation. Photograph by the author.

3. WHAT WE LEAVE BEHIND

ON THE FALL equinox of 2013, at the tail end of monsoon season, dense, dark clouds rumbled into a quadrant of northern New Mexico's pale-blue sky. Above the two- and three-story ruins at Chaco Culture National Historical Park, the skies were still bright. But the clouds, followed by virga and then by rain, meant the long drive out—fourteen of the twenty miles back to the highway are unpaved—would be a tricky one for me and my daughter.

Until then, we just watched the sky. And walked among the remains of the people who lived in the San Juan Basin of northwestern New Mexico during ancient times. Between AD 850 and 1250, thousands of people lived in Chaco Canyon. They built deep, round kivas underground and monumentally tall structures, with sandstone pieces so elaborately chinked and fitted together that mud mortar has kept them standing for centuries.

In his book *Anasazi America*, University of New Mexico archaeologist David Stuart wrote of how the Chacoan people inhabited forty thousand square miles of the Four Corners region. He explained that the "vast and powerful alliance" between the ten to twenty thousand small farming villages and almost one hundred district towns or "great houses" was reinforced by economic and religious ties. Hundreds of miles of well-worn ancient pathways, as wide as roads, are still visible from the sky.

The people who lived here also celebrated and mourned. They

fell in love, argued with one another, and stared up at the stars at night. Their feet touched the warm sandstone. They drew water from a running river and watched sticks burn to coals in their cooking fires.

Like now, there were good days and bad days—until the drought of 1090, when the bad days began outnumbering the good. The farming districts around Chaco—which supplied the great houses and also had the highest child-mortality rates—began faltering. By the late 1000s, wrote Stuart, firewood, clean water, game, and wild plant food had all become scarce.

It had taken the pre-Puebloan people more than seven hundred years to "lay the agricultural, organizational, and technological groundwork" that created the Chacoan culture, according to Stuart. The classic period, meanwhile, lasted only two hundred years—and collapsed "spectacularly" within forty.

On that fall day my daughter Lillie and I rambled along a trail near the base of a tall sandstone cliff. Near the ruin of Pueblo Bonito, the iconic structure that most people snap pictures of when visiting the park, seven or eight birds delighted in the waves of air from the stormy sky. They dove from the top of the cliff, twirled around one another, and pulled up just above the ground. They were turkey vultures, the vee of their wings distinct against the sky. I'd never seen those big, serious birds act that way, and as we continued walking toward the great house, we stepped into their game. They flew so close above us we heard the air through the feathers of their wings. Carrion eaters, the vultures are good reminders of mortality.

The clouds darkened the remainder of the sky. And the turkey vultures retired to a cottonwood in the wash nearby. We left and made it home, grateful for the rain that blessed our own hard drought.

But the same thing that delivered us to the canyon—my gasoline-powered pickup truck whose four-wheel drive also got us out that day—contributes to the destruction of cultural artifacts and threatens the future of the people who still live in small, tight-knit communities throughout the San Juan Basin in north-western New Mexico.

———

IN 2014 THERE were twenty-two thousand natural-gas wells in the region, some dating back to the mid-twentieth century. And as companies sought to glean oil from the tightly packed Mancos Shale thousands of feet below Earth's surface, the federal government started approving oil wells, too. If you've followed Highway 550 and driven north of Cuba, you've seen the trucks and rigs, the newer wells flaring thirty- or fifty-foot flames into the air.

Widespread natural-gas drilling started in the San Juan Basin in the 1940s, and today the landscape is dotted with tens of thousands of wells, most of which have been hydraulically fractured. In the 1980s companies also started drilling into coal seams to remove the natural gas trapped there.

By 2008, thanks to new and more efficient ways of getting at natural gas, wells in the San Juan Basin were churning out more of the product than any other region in the United States. Two years later, though, the market was glutted. Production was so high that prices plummeted and industry started hemorrhaging jobs. In 2010, five thousand jobs disappeared from the San Juan Basin.

In recent years, new technology allowed companies to extract higher-value oil trapped within tight shale formations. And that's what was driving a push toward new approvals of wells—hundreds proposed in the short-term—in northwestern New Mexico.

On a clear October morning in 2014, I joined a handful of reporters and then-councilman Mark Martinez, of the Pueblo of Zuni, to fly above Chaco Canyon in a six-seater Cessna 210. We took off from the Farmington airport and headed over the San Juan River and the Navajo Nation's thousands of acres of irrigated lands. In the distance the volcanic neck of Shiprock was visible, as were the lumpy heft and steaming towers of a coal-fired power plant. Banking away from the city, we headed south. Bruce

Gordon, the pilot and executive director of EcoFlight, pointed out the remains of ancient roads and buildings.

Driving on Highway 550 through the San Juan Basin, the desert scrub seems to stretch into nothingness. From the air it's possible to see everything else that's out there. Hogans and trailers at the ends of long, straight roads. Round corrals adjacent to a few venerated green cottonwoods surrounded by brown. The washes and arroyos that rarely merit a second glance from ground level are from the sky squiggly and magnificent. The badlands are rust and orange, white, gray, and black. Then the well pads and tanks come into view. The drilling rigs and waste ponds. From above, it looks as though someone fired a giant paintball gun. Globs of industrial development are strung together with roads like cobwebs.

For centuries this landscape has yielded what people needed. Once it was corn and beans, clean water, sandstone and timber. And in the past half-century, while farmers pull water from the San Juan River to irrigate thousands of desert acres, we also force the land to surrender coal and uranium, oil, and gas. My forehead pressed against the glass of the Cessna, I wondered how long this civilization will last.

There are also coal-fired power plants in this region, though as coal has become less economically attractive, electric utilities have also acknowledged the environmental benefits of closing plants. Near Waterflow, New Mexico, the San Juan Generating Plant is slated for eventual closure. And in November 2019 the Navajo Generating Station on the Arizona/Utah line closed.

But oil and gas, still booming, is harder to control.

After landing back in Farmington, Martinez explained what the flight meant to him. He'd been told that another 1,500 oil wells could be drilled in the basin within the next fifteen to twenty years. People at Zuni, particularly the religious people, worry about development approaching the ruins. Chaco is sacred to not just his Zuni people but to many Native cultures, he said. And for Native people, this place isn't an escape. It's a return.

When non-Native people visit the park, he imagines them

wondering about the people who lived there. They seem to miss the people who still live here, their descendants. "These are our lands. This is our heritage. That is our culture. We still walk upon the lands," he said.

"We call that our spiritual place and part of our umbilical cord to our migration route. So that's very important to us, and we've never left those homelands. They're part of our spirits," he said. "Our ancestors still live upon the lands. Every time we visit, we do offerings to greet them. Even though we don't see them, they're still in existence."

The overflight gave him a view of what's at stake. With about 20,000 wells already sunk, and 1,500 more planned for the next fifteen to twenty years, he was worried. And the development will definitely concern religious leaders back home at Zuni, he said.

But as industry refuses to relent its grip on the land and what's beneath it, communities, environmental activists, archaeologists, and some state and federal lawmakers fight back against the BLM's issuances of oil leases even outside the national park's boundary. "It's also about the larger landscape," said Samantha Ruscavage-Barz, who had a seventeen-year career as an archaeologist before attending law school and becoming an attorney for WildEarth Guardians. The fifty-three-square-mile area currently within the park's boundaries was the center of a larger community, she said, that extended as far north as Mesa Verde in Colorado and to the south of Interstate 40. All of those sites together tell a story.

Allowing development up to the boundaries of the park creates something "almost like a little playground," she said. "You don't get that sense that there was a living community occupying that landscape, and the descendants of the community are still there." Today, that living community includes the Diné (or Navajo) people, who have homes on the reservation and on allotments, private lands deeded to individual families in the nineteenth century.

———

ON A MID-DECEMBER morning in 2014, Lori Goodman arrived outside the Counselor Chapter House on the Navajo Nation. A handful of people, mostly women, stood in the dirt lot, soaking in the New Mexico sun on a day that probably should be a dozen degrees colder. "Quite a sight, huh?" asked Goodman, who volunteers with the nonprofit organization Diné CARE. She's referring to the oil rigs we passed to get here. In this area, the BLM just approved about one hundred new oil wells, some of which popped up alongside the unpaved roads that run through Counselor and south of the highway toward Ojo Encino.

Etta Arviso is originally from Nageezi, and her grandfather grew up near Chaco Canyon. "I know there were sacrifices that were made for me," she said. She's a descendant of the Long Walk, the forced removal of Navajo people from Arizona and northwestern New Mexico in the nineteenth century. In 1864 the US Army marched almost nine thousand Navajo men, women, and children to the Bosque Redondo Reservation on Fort Sumner in eastern New Mexico. In 1868 the Navajo Nation and the US government signed a treaty allowing the people to begin their "long walk" home.

"My grandma was on the Long Walk when she was only three years old, as a prisoner of war under the United States flag," said Arviso, who also works with Navajo Code Talkers. "These companies need to know that's who they're disrespecting: Men who used the Diné language to save the world," she said. "They need to step back and back off and know they are visitors." It's not just the drills and flares and waste ponds that upset Arviso. It's also the way the roads get ripped up: "If these roads look like they need to be improved, they better doggone well sit down and negotiate with the county, or the BIA, or whoever allowed all these signatures, allowed all this to be happening."

Many allottees who signed contracts allowing companies to drill next to their homes or on their grazing lands didn't understand what they were signing away, she said. "People who signed contracts, some of them couldn't read, some of them couldn't understand . . . they were told, 'Just sign the paper. We're raffling this off, sign your name, you're going to get paid.'" She's mad that Navajo leaders didn't stand up to the companies and that the federal agencies, like BLM, have allowed so much development where Diné people live.

"How in the world can the BLM just allow this to happen?" she asked. "The federal government no longer needs to say, 'I have this in trust.'" (Tribal lands are held "in trust" by the federal government in the United States.) "They need to give us back the deed if they're not going to have people they hire to protect us, so we can take care of it."

Later we drove deeper into the reservation and then climbed atop a pine-covered hill to look out at the landscape. Another Diné woman, Sarah Jane White, walked with us, too, pointing out footprints in the sand to her four-year-old grandson. He crouched to look at the prints of a horse, and then a cow. Each time, he's delighted. We walked past juniper trees and an old sweat lodge. Across a sandstone arroyo and up a hill.

A small ranch, marked by a windmill and a few barking dogs, spread out below. The air was warm and smelled like sage or pine, depending on the direction of the wind.

"Right now, there's healthy people living here," said White. "The air is fresh. It's clean." White and her relatives are "allottees," Navajo people living on lands deeded to them by the federal government.

Then, from behind the mesa that flanks the ranch, we spotted a flame. Then another farther to the right. Below, there's a well pad with huge, green storage tanks. Turning again, we saw another flare. People like White who live here seem surprised by how fast things have changed. "When they're done sucking everything out, everybody's going to pack up and leave and leave their trash behind," she said. "Nobody's going to clean it up. That's what bothers me."

At the top of the hill, White looked out across the landscape where she could see four different wells in the distance. "I see the landscape looks really beautiful, but when you see all these oil tanks and fields, that's not beautiful," she said. "The flare, that doesn't look good at all. And if we don't stop this, it's going to be all over the place."

Here, on the eastern edge of the Navajo Nation, there are about nine hundred thousand acres of allotments—and this is where most of the current oil drilling and proposed leasing is occurring. For a one-time payment, sometimes as much as $100,000, hundreds of individual families have signed agreements with drilling companies. Each well pad has its own road, waste pond, and tanks. During drilling, pickup trucks and semitrucks run up and down the roads twenty-four hours a day. And with the money come lasting effects. Oftentimes, Goodman explained, people don't understand what they're signing, and they don't understand what's going to happen on their lands. That's also a one-time payment—even if the well runs for decades.

Many Diné people who live here are upset that the roads leading to their homes are being ripped up by semitrucks. They're afraid of the chemicals companies use to hydraulically fracture— or frack—wells. Pioneered by the energy company Halliburton in the 1940s, hydraulic fracturing, or fracking, is the process of using sand, water, and a cocktail of chemicals like formaldehyde, benzene, and hydrochloric acid to open up small fissures belowground to better release more natural gas or oil. Additionally, Goodman said that people don't know what's coming out of the flares. And they worry about blowouts and accidents that happen far—very far—from emergency services.

"People that we've been meeting with, they're telling us how this intense development has really interfered with their lives and livelihood," said Goodman on the walk back down the hill. "Numerous roads have been built, just in this rush to get the oil out. And so regular roads that were just to people's homes are now truck roads. The amount of traffic going by—it's scary. We were driving in a small car, and these huge semitrucks were passing us, and we kept pulling off the road because we're little and they're huge."

One woman, who later didn't want her name in the news, told

me, "Especially at night, it's enough to make you just cry. One of the ladies [said it] looks like a war zone. It's just completely lit up. All you see is flames everywhere, you smell that gas, that burning, it's just ugly."

She knows that the wells mean money: lots of cash for people working in the fields or depending on land ownership and jurisdiction, hundreds of millions of dollars for the state of New Mexico, the federal government, or the Navajo Nation tribal government, and millions more in profits for the oil companies. But she's angry that Navajo people are living with the trucks and the flares, the noise and the fear. "I think Indigenous people, Navajos, we've been pushed around enough. We were forced to live on land no one wanted, [and] now everyone wants it because we're full of natural resources," she said. "It's not right. And so, leave it where it's at. Leave it where it's at. That's what I say."

White said Navajo people live where they are born. "Like, if I was born here," she said, pointing to the ground between her feet, "I would live here. And I would die here. And I would want to be buried here. You don't leave your homeland." That's why they are fighting, she said. Because what happens now will still matter to the children born here in a century.

———

AFTER EVERYONE ELSE heads home from the Farmington airport, Mike Eisenfeld, energy coordinator for the nonprofit San Juan Citizens Alliance, and I stick around, claiming a picnic table with a view of the tarmac while the wind beats the metal weight of a flag against the pole. The *ting, ting, ting* is like a metronome pacing out our conversation. In addition to the tens of thousands of wells, the Four Corners has coal mines and coal-fired power plants. And whether it's from coal, oil, gas, or uranium development, the air and water pollution in the San Juan Basin serve as proof, for Eisenfeld, of "our continued inability to think about ways to create electricity other than from burning stuff."

It's a shame, he said, that the Four Corners is considered a sacrifice zone—a place where America's pursuit of energy independence always comes home. "I don't think, until you live in a community like this—where there are well pads in every park and well pads in your communities and well pads next to your schools, and compressor stations and processing and central delivery points," he said. "I think in my mind, I thought that like natural gas just came out of the ground clean and wasn't really that big of a deal," he said. Living here changed all of that for him. I've known Eisenfeld for years, and he's never once relented. Never once given up on his adopted community, a place he's raised his family and built relationships.

"My experience here is this is a tremendous living community, full of history and vitality, and it's just a shame that some of these monikers have stuck," he said. "The current president's [Obama's] 'all of everything' energy policy is not well-thought-out at all—because the areas that have been impacted for years and years by oil and gas and coal and uranium and other schemes and shenanigans are going to just be constantly hit with the idea that we need to be relied on for more and more and more."

Eisenfeld burned with anger about the wells near Lybrook on the eastern Navajo Nation that flare off nitrogen and natural gas or methane. The wells and all their accompanying tanks, trucks, roads, and pipes have industrialized the landscape. Wells churn across the street from Lybrook Elementary School and up against people's homes and hogans. They lay waste to peoples' lives and health every day—and they also contribute greenhouse gases to the atmosphere. But it's what those flares and leaks represent that make him angriest. It's so wasteful, so poorly conceived, and so heartbreaking, he said, that as a society we're so willing to squander things with environmental problems and economic value. The methane that's flared or leaked into the atmosphere is natural gas, after all—the same natural gas we use to heat our homes and boil our pasta.

He keeps defending this ground, working as an ally and an

accomplice. But he's frustrated that he doesn't know what to do about how humans have fundamentally altered our natural systems. He's not just talking about the landscape of an ancient civilization pocked with wells and mines. He's referring to the hand New Mexicans have had in altering the planet's climate. Lest we forget, in 2014 NASA revealed that methane anomaly in the Four Corners region.

During the second century, when the rains and snows stopped falling with regularity in the Four Corners, I imagine that people felt some combination of hope and helplessness. Hope that next season would be better, that the stores of food could be renewed again after this one tough year. Hope that gods would answer prayers. Then helplessness as more children died. As water became scarce and wild game disappeared. As crops failed to survive the dry summer and the walk for firewood became too long. When our downfall comes—as the Southwest continues to warm, as surface water supplies decline, and as the evidence of our actions is visible even from space—we won't leave behind remains as beautiful as those at Chaco. We'll leave behind seeping gases and rusting pump jacks, poisoned wells and cracking pavement.

A few months later Eisenfeld and I met up again. We drove the back roads around Counselor, trying to grasp the scale of the new oil development. I followed behind him as he rushed to see one more well pad, one more rig, before darkness descended completely.

After sunset we parked at a cleared, flat spot. It seemed like an old natural-gas well pad, but I couldn't quite make out the infrastructure in the darkness. He stopped here to watch the flare from an oil well. It was about a quarter-mile away, but we yelled to hear one another over its roar. Then Eisenfeld hopped back in his truck and took off again.

As he pulled away, a great horned owl swept up, landing on a pole atop an elevated metal tank. My heart thumped from the surprise of the bird's giant wings. I wanted to watch this owl forever. Or as long as my eyes could see through the dark. But I had to get moving—I don't know the route back to the highway. As Eisenfeld's taillights disappeared down the gravel road and around a corner, I remembered what he'd said to me a few months earlier. I remembered his urgency.

"I'm fifty-two, and I'm kind of going, 'All right, what's going to happen in my lifetime? And what kind of planet are we leaving our kids?'"

———

Over the coming years the battle over these lands intensified. The Western Environmental Law Center, representing environment and community groups like Diné CARE and San Juan Citizens Alliance, sued the federal government over the leases. In 2017 the Navajo Nation asked the BLM to enact a moratorium on multistage fracking, horizontal drilling, lease sales, and permit approvals until a new resource-management plan is completed and an Environmental Impact Statement finalized. In their letter to the federal agency, Navajo Nation President Russell Begaye and Vice President Jonathan Nez wrote they were concerned that drilling activities were "interrupting the daily lives of Navajo people who live in the Navajo Nation Chapters such as Counselor, Nageezi, Torreon and Ojo Encino." Shortly after that the two Navajo officials met with the All Pueblo Council of Governors. In a historic meeting that brought together the Navajo Nation and nineteen New Mexico pueblos, the tribal governments called for protection of the areas around Chaco. They also created a working group to address the issues.

During the 2017 state legislative session, freshman state representative Derrick Lente watched one of his first initiatives turn into a showdown on the House floor. Earlier in the session Lente's memorial to protect cultural and historical sites near Chaco Canyon received bipartisan support and passed through the House State Government, Indian and Veterans' Affairs Committee unanimously. Something had changed, though. By the time it reached the House floor, the Democrat's memorial had triggered uncertainty and skepticism from Republicans.

That's because there was an elephant lurking in the room, said Lente, who is from the Pueblo of Sandia. "It's the f-word," he said in an interview afterward. "Fracking."

Republicans opposed the memorial, which passed 31–28 on a party-line vote.

Citing the historical, cultural, and economic importance of the hundreds of sacred and archaeological sites in northwestern New Mexico, the memorial encouraged the BLM and the Bureau of Indian Affairs to halt new leases and drilling permits until there has been sufficient tribal consultation and a new resource-management plan has been completed.

Beyond the boundaries of Chaco Culture National Historic Park, where sandstone villages, kivas, and Great Houses are preserved, hundreds more sites radiate out across the landscape. Drilling opponents say the current plan, written in 2003, is out of date and does not consider the cumulative impacts that hundreds of new shale-oil wells will have on the sites or the landscape and its communities.

In an earlier memorial, Lente and Rep. Patricia Roybal Caballero, D-Bernalillo, had requested that the BLM consider a temporary moratorium on all "fracking-related" leases and permits in the Greater Chaco area until the agency completed its plan. The House Energy, Environment and Natural Resources Committee tabled that memorial.

The new memorial, Lente said, "didn't take such a deep stab." Instead, he said, he wanted to encourage the BLM to follow the federal laws requiring consultation with tribes and to truly investigate what impacts the oil and gas industry is having on the area. "This was not about disenfranchising an industry or trying to impact New Mexico's dwindling financial status," Lente said of the memorial. "It truly is an outreach, to try and bring people together."

Lente said that when he was campaigning in the district, some

Navajo people living near the development along Highway 550 said they felt like they didn't have a voice. People living near the new wells—many of which are clearly visible from the highway—don't know what's happening underground or to the air around them, he said. And when people don't have all the data, or feel like no one is listening to them, they start speculating about what could happen.

Tensions are growing, Lente said in 2017, between people in the communities who support oil and gas and those who oppose it. Local people's feelings are far from uniform when it comes to the drilling. "It's causing a lot of in-fighting," he said.

That's why he and others would like development halted until the agency knows exactly how drilling is impacting communities, landscapes, water, and the environment. Once there is concrete data, he said, there can be discussions around the table involving everyone. In the end, he said, those studies could show that what tribes feared is actually safe, or they could confirm that drilling is causing serious problems. But at least everyone would know and be able to move forward from there.

Diné activist Daniel Tso pays close attention to the development. In 2017 he explained that he noticed new wells off Highway 550 along the road to Chaco Culture National Historic Park, as well as continued drilling across the highway from Lybrook Elementary School. He also watched operators bring in dozens of temporary storage tanks to a WPX Energy oil production site near Nageezi, where in the summer of 2017 tanks ignited in a fireball. That same year, the New Mexico Senate confirmed retired WPX Vice President for San Juan Basin Operations Ken McQueen as the secretary of the state's Energy, Minerals and Natural Resources Department. (Early in 2018, WPX Energy sold its San Juan oil play to another company for $700 million. That, on top of its other divestments in the basin, meant the company shifted entirely away from northwestern New Mexico, choosing instead to focus on more lucrative fields in the Permian Basin and in North Dakota.)

"The fact is that after these facilities are put in, there is rancid, foul-smelling air," said Tso, who has also served on the Navajo Tribal Council as a representative for Torreon Chapter. He described how first the rig comes in to drill the well. "There's the constant noise of the drill, which can last seven days to two weeks, when you might as well be at the Sunport [Albuquerque's airport], listening to jet engines."

Once a well has been drilled, trucks haul in materials like cement and steel pipe, and then water, sand, and the chemical mixtures for fracking. "Then, once the fracking operation is completed, they have to truck out the produced water, and truck out the oil," he said.

All that heavy traffic ruins the gravel and two-track roads that wind between communities on and near the eastern Navajo Nation. These are roads that buses use to take students to school, said Tso. The four-wheel drive pickups and semitrucks grind deep ruts into the roads, he said—sometimes a foot or two deep. Then, after they've been graded to smooth out the ruts, Tso said the roads end up two to three feet lower than the surrounding land. They collect water, which further erodes them. "You might as well have an irrigation ditch, but who's going to take ownership? Who's going to install culverts, or elevate the roads so water doesn't puddle up in the middle?" Tso asked.

Tso said he's grateful to Lente, who he said was "true to his word" in standing up for the people living in places like Lybrook, Counselor, Nageezi, and Ojo Encino. Next, Tso would like to see the state's congressional delegation take more of an interest in the area—to help find resources to fix the roads and encourage BLM and BIA to take their Indian Trust responsibilities seriously. He would also like to see baseline studies of the water, soil, and air performed in places where development isn't yet occurring. That way people can know exactly how development, as it moves forward, affects people and the environment.

Tso said that people sometimes look around and ask where exactly the "Greater Chaco Landscape" is or how people came up with the name. "It's not a point on a map. It's not designated by longitude and latitude," he said. "The whole landscape is sacred."

The signs of ancient life are everywhere in New Mexico—from broken pottery pieces on the desert floor to the Puebloan design in our modern buildings and the communities that still rely on the wisdom passed down from ancestors—and they are perhaps most obvious in northwestern New Mexico. There, the landscape yields Great Houses and kivas, scatters of stone tools and black-on-white painted pottery. And close-knit communities still rely on the landscape they've lived upon for centuries. The signs of what's coming are obvious here, too. The boom and bust of the energy cycle continues to play out, and while the booms can bring economic prosperity to some people, the busts always leave behind industrialized landscapes and communities unsure of what their next steps might be. And both the booms and the busts show us what it looks like to squander lands, resources, and relationships.

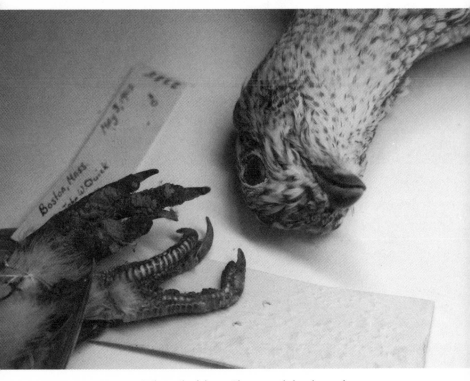

FIGURE 5. Aplomado falcon. Photograph by the author.

4. MOURNING A MOUNTAIN

ON A CRISP, late fall morning in 2015, Larry Rasmussen, a Lutheran lay minister, and I sat side by side at his kitchen table in Santa Fe. Sometimes we looked at one another, but mostly we both looked out the picture window at a sky cerulean blue and at songbirds flitting about the backyard—from bird feeder to fence post to bird feeder and *hop, hop,* around and around again.

More than a decade earlier Rasmussen retired from New York City's Union Theological Seminary. He and his wife now live off Old Pecos Trail in a wooded part of Santa Fe. From this window, they can watch a forest in transition.

"Forest in transition." That's how we talk now about mountainsides full of trees that are dying. Trees that smell like vanilla when you press your face close in warm sunlight. Trees whose sticks and split trunks have been fed for generations into cooking fires and wood stoves. The juniper and piñon wood smoke that wafts through cities and towns today is the same that's filled the air since Pueblo people settled along the Rio Grande centuries ago.

But as the world keeps warming, not all those species will survive. Scientists keep saying this, and even drive-by tourists can see the waves of young oak trees spreading across burned slopes that conifers once dominated. I can't help but wonder how differently the land will smell when our children's children inhale; I can't help but wonder if they'll ache for something they don't know is missing.

"What do you do with your grief? What do you do with your loss?" Rasmussen asked the questions for me. "What do we do with species going extinct? What do we do with nude mountains?" We both looked out the window again. He's talking about a wildfire that ripped through the Jemez Mountains a few years earlier.

At about one in the afternoon on June 26, 2011, the wind blew an aspen tree into a power line running through the Jemez Mountains. At the time Las Conchas Fire started, it was 90°F in the forest with just 6 percent humidity. Stage III fire restrictions were in place already due to the fire danger—meaning backcountry camping, hiking, and recreating was prohibited—which officials credit with saving lives, since evacuations would have been impossible given how fast the fire moved.

Through their Santa Fe kitchen window, Rasmussen and his wife, Nyla, watched the mountains burn. At night the hulk of the mountains—normally just a crouching, dark blankness swallowing the pinpoints of stars—flamed orange.

It's not that southwestern forests aren't supposed to burn. But this fire was different. In its first fourteen hours, Las Conchas burned through about one acre a second.

Close your eyes and count to ten. During that time, flames devoured ten acres. Think of the ponderosa pines and Steller's jay nests, the fox burrows and salamanders obliterated in the time it takes to draw and exhale two breaths.

All told, the fire burned 156,000 acres in the summer of 2011.

That was the state's largest wildfire in recorded history. Until the next year when the Whitewater-Baldy Fire burned nearly twice that amount, 297,000 acres, of the Gila National Forest.

In our warmer world, ponderosa pine stands don't necessarily come back. Instead, they give way to Gambel oak and New Mexico locust. Piñon and juniper trees die off, their hard carcasses thronged by wavy leaf oak and mountain mahogany.

In 2016 scientists from Los Alamos National Laboratory, Columbia University, the University of New Mexico, and a pile of federal agencies and universities published a paper in the peer-reviewed journal *Nature Climate Change* that predicted the disappearance of the Southwest's pine-juniper forests. It followed a 2010 paper about how warming is affecting southwestern forests and another in 2012 that found that 20 percent of the region's forests had been affected by beetle infestations and high-severity wildfires since the late 1990s.

The 2016 study found that earlier mortality estimates were too low—and it projected that by 2100, pine-juniper forests in the southwestern United States will be gone, and more than half the evergreen trees in the northern hemisphere will have died.

That large-scale die-off will have global impacts.

As the authors pointed out, forests absorb carbon. If tree die-offs continue at this rapid rate—exacerbated by warming temperatures and more extreme drought events—that carbon absorption function could be diminished enough over the next century to further accelerate warming.

In the Sandia Mountains, which frame the east side of Albuquerque, there are thousands of acres of dead conifer trees. In the fall of 2015, when US Forest Service silviculturist Shawn Martin took me up the west face of the Sandias, there were roughly nine thousand acres of dead trees—Douglas firs killed by tussock moths and engraver beetles and lower-elevation trees decimated by infestations of piñon ips beetles.

Insect outbreaks are a massive problem in southwestern forests, where drought and overgrowth have weakened millions of acres of trees. Two years earlier the Forest Service conducted aerial surveys over twenty-one million acres, and while the number of acres affected by new insect-related die-offs was down from the previous year, it was still high: Ponderosa-type bark beetles defoliated 70,110 acres in New Mexico; mixed conifer-type bark beetles, 66,620 acres; spruce-fir type bark beetles, 21,550 acres; and western spruce budworms killed more than 300,000 acres.

This isn't just happening in northern New Mexico. In the Gila National Forest, nearly three years after the Silver Fire raged through the forest in southwestern New Mexico, dead trees stood like matchsticks along Emory Pass. Their charred skeletons cast long shadows down the slopes, and I found myself holding my breath as I drove along the highway.

Long before lightning ignited the dry, overgrown forest near the tiny town of Kingston in June 2013, foresters knew the area was at risk. In some ways the Silver Fire exemplified what happens when the legacy of nineteenth-century management practices meets twenty-first-century warming. You get weak, overcrowded forests that double as kindling in which fires can spread quickly.

Historically, fire played an important role on the landscape. Many of New Mexico's mountain-tree species evolved with fire; ponderosa pines, for example, thrive on regular cycles of low-intensity fires. But beginning in the late nineteenth century and lasting into the 1970s, a national policy of fire prevention and suppression aimed to protect commercial timber stands and newly designated "forest reserves" alike. Under that model, all fires were perceived as bad, and foresters were prompted to pounce on and douse even the smallest of fires.

Such practices changed the character of many forests, including the Gila near Kingston, which grew dense with shrubs and small-diameter trees. In 2004 insects attacked conifers weakened by drought, killing nearly all the white firs. Meanwhile, steep terrain and a lack of roads prevented crews from clearing the dead trees.

So when lightning struck that hot, dry, and windy spring, the Silver Fire feasted on the forest's ample fuel.

For more than a month firefighters tried to keep flames within federal boundaries and protect the adjacent private lands. "We lost the whole south end of that mountain range," said Gabe Partido, timber program manager for the 3.3-million-acre Gila National Forest. "We lost probably 80 to 90 percent of the mixed conifers up there."

Just in the span of three years, the Gila experienced three huge fires: In 2011 the Wallow Fire, which started in Arizona, consumed more than 500,000 acres. The Whitewater-Baldy Fire, the state's largest wildfire on record, devoured almost 300,000 acres in 2012. Then the Silver Fire burned 138,000 acres.

Up in the Santa Fe National Forest, where Las Conchas burned,

Dennis Carril, the forest's fuels specialist, told me that during the 1980s about 2,500 acres burned. Then, between 2010 and 2014, a total of 150,000 acres burned within the 1.6-million-acre footprint of the national forest. That doesn't include fires outside their boundaries: For example, Las Conchas burned a total of 156,000 acres, but roughly half of that burn was on adjacent lands overseen by the National Park Service, the state, private landowners, and four pueblos.

"Fire is inevitable," said Carril, who advocates for paring back overgrown forests, reintroducing fire through prescribed burns, and continuing to reverse that old model of putting out even naturally caused fires. But, he said, fires like Las Conchas aren't normal.

That "uncharacteristic" fire destroyed the forest's canopy and its seed sources. The impacts of those fires can last for years, even generations.

Without roots in place to slow water, floods rip apart arroyos and stream beds, damaging downstream communities, degrading water quality, and destroying wildlife habitat. High-intensity fires cook the soils, and drier, hotter conditions can prevent trees from returning. In New Mexico that means brush is replacing piñon pine and juniper forests; locust, aspen, and oak are taking over lands that were once conifer forests.

How do we mourn nude mountains?

At my father's funeral in 2013, I realized how much I missed those ceremonies we wrap ourselves in during times of grief or confusion. At the time, I was thirty-nine years old and had lived more than two thousand miles from my parents for nearly twenty years. But I'd never been to a funeral without my father. I'd never watched the casket enter a church or sat through a eulogy without placing my hand inside his. Even if we'd argued on our way to the church, I'd position myself next to him during the service, then hang around him and his pals—fellow cops, oftentimes—during the meal afterward. They'd stand around outside, tug loose their ties, and insult one another. Another ritual.

Is there a way to mourn a river, too?

Two decades ago, two Fish and Wildlife Service biologists sat behind their office in Albuquerque. Surrounded by boats, trailers, tanks, waders, and all the accoutrements of fisheries biologists, they realized the fate of an entire species rested with them. Historically, the silvery minnow had lived throughout the Rio Grande and its tributary, the Pecos River—about 2,400 miles of habitat. By the time it was protected under the Endangered Species Act, the minnow survived only in a 174-mile stretch of the Rio Grande.

Like many species, the minnow's demise was linked to human cleverness. In the twentieth century, people had tamed, tapped, and engineered the Pecos and the Rio Grande to prevent flooding and deliver water. Their work was so effective that watersheds and riverbeds bowed before miraculous engineering feats. Ecosystems changed. Species disappeared.

Feeling the tug of moral responsibility for an entire species, the two biologists convinced their bosses to let them try something different. They started collecting eggs from the river and then raising the fish in a hatchery. They did that because New Mexico's largest river—dammed and managed and dried—could no longer keep a two-inch-long fish from going extinct.

Speaking years after he left the Fish and Wildlife Service and New Mexico, fisheries biologist Christopher Hoagstrom still gets riled up about the Rio Grande. "New Mexico should be proud of its rivers," he said. He was talking about not just the Rio Grande, but also the Pecos River in the eastern part of the state and the Gila, in southwestern New Mexico. "And we should ask ourselves, 'What happens to areas, long term, without their rivers?'"

That answer can be seen throughout the Western United States, he said. From Fort Stockton, Texas, where oil drilling and roadside trash clutter former grazing lands, to the Owens River Valley of California, a lush landscape that withered and dried after its water was siphoned off to Los Angeles. Pumping groundwater can work to sustain places—for a while. But even those waters are overused. And those systems can often take tens of thousands of years to be replenished.

Drought isn't like a landslide. It doesn't slough off a mountainside, destroying everything all at once. Rather, it ekes away at life over the years. Puny snowpacks don't refill reservoirs. Dry soils are stirred up and blown away by spring winds. Deciding to pack it in, one farmer, then another, sells off water rights to a city. An old stand of cottonwood trees isn't replaced by saplings. A species of fish goes extinct.

Then one day a new generation wakes up and doesn't remember that a river once flowed through their community. A dry riverbed is no longer something unusual. Rather, it's just another dusty rut in the landscape of memory.

"Human cultures rise and fall on water—and we're a part of that," Hoagstrom said. "People think of the Rio Grande as something that's separate from their lives or from what they're doing. But it's an indicator, and the drying is a foreshadowing. If we dry the river, it shows we're using water unsustainably."

What do you do with your grief?

At the end of my father's wake, no one rushed me. Friends and family and police officers looked away. They were ready to walk with the casket from the wake to the church. But I wasn't ready to leave the casket. Not ready for the lid to seal shut. I knelt, pressed my head against the cool metal, and then recoiled at the smell of rot and chemicals. By the time the pallbearers carried the flag-draped coffin up the steps and into the church, I'd lost track of whether he was head or feet first. Incense. Words. Stained-glass windows and gold on the altar. From the church's front pew, I stumbled to the altar in heels and a dark-colored dress, then choked through a reading. At the cemetery uniformed men and women fired the requisite number of rifle shots. Someone handed my mother the folded American flag. The ancient driver of the hearse blew into a trumpet. In the back of the limousine, I held my brother's hand. For months afterward, I thought about decomposition. I imagined that the blue eyes I inherited from my father no longer existed within his face. The lips and mouth that looked like mine were surely gone. His hand must be desiccated. I placed the rifle shells in a wooden box. But I still don't know where to put my grief.

Back in Albuquerque I visited with a biology professor at the Museum of Southwestern Biology at the University of New Mexico. I'm curious about the Latin American hummingbirds whose evolution he studies. After talk of jungle mountains, lungs, and tiny hearts, Christopher Witt walked me to the back of the museum and slid open a wide, shallow drawer.

There were passenger pigeons, Carolina parakeets, Ivory-billed woodpeckers, dusky seaside sparrows, and Eskimo curlew. Nowhere do they have descendants, preening or molting, blinking their eyes against the rain or singing a morning song. These species are all extinct.

Another drawer held a tray of threatened Mexican spotted owls, their downy, white eyelids closed. There were also boxes of Northern aplomado falcons, which were nearly extirpated from the United States by the 1950s. I peered at the tag tied to one falcon's foot: "*Falco femoralis septentrionalis*. Fort Bayard, New Mexico." Collected by F. Stephens on August 8, 1875.

My eyes watered, and I looked down in shame. I jotted notes I couldn't actually read. How do you mourn a drawer full of birds?

My father had been dead two years when I returned to Connecticut for three summer weeks. After living for almost twenty years in the Southwest, I scoff at the green. The trees crowd out my necessary horizon; I anticipate ticks. But I love getting to know my cousin's children. I'm secretly pleased when my daughter, brazen like I never was, mouths off to me then runs into the backyard with her kin. In the woods beyond my mother's house, my childhood fort is gone. It's not that someone scattered the brush and stones and pieces of metal I'd collected and piled. The whole hillside is gone. More than a decade earlier, a developer cut down the trees and blasted the mica-flecked bedrock. When he ran out of money, he abandoned the subdivision and left behind a backhoe and piles of construction waste. Now, water fills empty foundations. There are cattails and frogs that disappear in a croaky ripple. In time it turned into a dumping ground for other people, too. An old Christmas tree lies across the road;

televisions and suitcases flank it. I poke assorted weird shit with a stick.

Beyond the mess, the forest is thicker than it was decades ago. There are foxes and turkeys, deer and species of birds I don't remember seeing when I ran around the woods as a kid, tracing the paths of stone walls, forever hopeful I'd discover a gateway into another world. That other world might be magical, it might be sinister. Either way, I was ready, pocket knife filched from my dad's collection.

During one of those last summer nights visiting my mom in Connecticut, I stared up at the stars around the edges of the trees and heard a sound that's familiar to me only from the West.

Coyotes. At first I thought I was dreaming. I held my breath and listened. The syncopated stutters carried across the forest from a hillslope away. Past the roadcut and power lines.

When I first started writing about climate change, David Gutzler said not to worry for the planet itself. It's humans who will suffer, he said. The earth will remain. Species will blink out, evolve, or emerge. Ecosystems will shift. The same geological processes that have been churning the earth for eons will still shape tomorrow's landscapes.

The selfish grief I felt over losing my father, and any chance he and I had for understanding one another or reconciling our differences and disagreements, has dissipated. In its place a fiercer love has emerged. For my daughter and my mother. My brother and his family. Dear friends and ready allies. For the world we still have. Looking back, I realize that I twisted nostalgia and hope into inaction and despair.

Perhaps healing requires grief and mourning. But both those emotions seem like indulgences, and there's no time for that. So instead I wonder about the coming world. Which trees will grow, which birds will have survived. The meaning of shifting cloud formations. I wonder what the world will smell like. Because now, even though it's been decades since I stopped looking for that gateway in earnest, the door to that new world has opened. And there's no going back.

FIGURE 6. Jemez Mountains. Photograph by the author.

5. CLIMATE OF FAITH

AFTER WILDFIRES LIKE Las Conchas, or any of the others that have ripped through New Mexico's forests in recent years, green returns to the landscape. But different species from the piñon and ponderosa we lost often grow back. As plant communities change, so do the species of birds and wildlife that rely upon them. And the relationships people have with the forest—to earn a livelihood, recreate, or find solace—change as well.

"Scientists can describe the boundaries that we dare not cross," said Larry Rasmussen, the Lutheran lay minister who lives in Santa Fe. Scientists can describe intensifying wildfires, droughts, disappearing glaciers, the extinction of species, and rising sea levels. They can predict and model. But data points and scientific graphs don't inspire people to change their behavior. That takes faith. And love. "It's the values that people want to live their lives by, or the love they feel for a place or for their family or for their friends that motivates them." Rasmussen, who authored a book called *Earth-Honoring Faith*, has been working toward an alliance of spirituality, social justice, and ecology since the late 1960s—and he believes action within religious organizations is finally reaching critical mass.

In 2015 Pope Francis issued an encyclical letter focused on the earth, the economy, and social justice. Popes have issued these letters regularly since the eighteenth century, but "On Care for Our

Common Home, Laudato Si" is different. The encyclical isn't what most people would expect from the Vatican. In his 120-page letter, the pope cites not only saints and the Catechism of the Catholic Church, but scientists, economists, and Indigenous leaders. He also calls forth the words of other religious traditions and leaders, including Ali Al-Khawas, a ninth-century Muslim mystic.

With his letter, Pope Francis issued an international call to action that reminds people that all life is interconnected. He called upon humans to recognize that everything we do affects all of life, from the smallest microorganism to the largest mammal, from the atmosphere that sustains us to the planet's most vulnerable men, women, and children. It is a call for structural economic changes—especially within nations like the United States that have amassed wealth by building industries that emit huge amounts of green-house gases. The pope believes these structural changes will ease demands upon the earth and protect the planet's poorest people—who are already experiencing the brunt of the impacts from climate change.

Rasmussen knows people will resist the messages within the encyclical. Since the beginnings of the Industrial Revolution, economies have thrived on natural-resource extraction, the burning of fossil fuels, and an unshakeable belief in continual growth and the importance of individual successes over the common good. "But the way we did it in the past is no longer viable," he said. "We can't keep working to keep the same economy that gave us climate change."

People can take individual steps and work toward larger policy changes in their cities and states, he thinks. Organizations like churches can install solar panels and cisterns the way United Church has. But people must change more deeply, internally, by rethinking how we connect with water, landscapes, and one another. "And it's pretty urgent that faith communities are involved," Rasmussen said. "It's their home turf."

By insisting that faith communities are critical to the success of the climate conundrum, Rasmussen is on to something important.

According to a 2015 Pew Research Center survey, 80 percent of Americans claim affiliation to some religious faith. Like philosophy and great literature, communities of faith traffic in grand narratives about the meaning of life, morality, and the place of humans within the universe. And it's not unheard of for those communities to step up when it comes to US social movements. Think about abolitionists prior to the Civil War and the civil-rights movement; many of these African American leaders came out of the black church tradition. And today some religious groups are taking a bigger role in the environmental movement.

In early September 2015 an evening storm brings wind, but no rain, to downtown Albuquerque. About a dozen people set locally grown food around a table and pass plates around, introducing themselves to one another while they eat.

Father Frank Quintana of the Blessed Oscar Romero Catholic Community has convened a four-week study group of the pope's encyclical. The papal letters are typically written to bishops and the faithful, explained Quintana, but Pope Francis wrote this for everyone, and it showed in the attendees—like non-Catholics and a lapsed-Catholic journalist who worries about churches bursting into flames if she passes too near. "God gave us creation as a gift, and instead, we have acted as if we are lord over creation," Quintana said. "That has had a disproportionate effect on the poor. In essence, we have robbed the poor. And [the pope] is calling on us to take care of each other and creation."

That message resonates with Javier Benavidez, then-executive director of Southwest Organizing Project, a nonprofit that focuses on social and environmental-justice issues. "New Mexico is like a microcosm for the world," said Benavidez, whose wife brings their youngest son, just newly born, to the meetings sometimes. "There is a lot of poverty, but it is resource rich. We provide cheap labor for industry, then suffer the pollution." He ticks off just a few events on the timeline of New Mexico's environmental history: The August 2015 Gold King Mine spill into the Animas River, which flows into New Mexico's San Juan River (that spill, from an abandoned mine in southern Colorado, is just one instance of a long-abandoned, privately operated mine contaminating the state's waters). The 1979 release of more than ninety-four million gallons of toxic wastewater and a thousand tons of radioactive tailings into the Rio Puerco. And the nuclear detonations at the Trinity site in central New Mexico in the 1940s. And while some people may shrug off the poverty, the pollution, and the economic dynamics by saying it's just the way things have always been done in New Mexico, Benavidez doesn't buy that. Not even for a moment. "There are interests that come here and now are actively pushing coal or thwarting clean energy," he said.

In the encyclical, Francis quoted the bishops of the Patagonia-Comahue region of Argentina: "We note that often the businesses which operate [in less developed countries, in ways they would not at home, where they raise capital] are multinationals. They do here what they would never do in developed countries or the so-called first world. Generally, after ceasing their activity and withdrawing, they leave behind human and environmental liabilities such as unemployment, abandoned towns, the depletion of natural reserves, deforestation, impoverishment of agriculture and local stock breeding, open pits, riven hills, polluted rivers and a handful of social works which are no longer sustainable."

The pope's message also resonated with the leaders of Juntos, a program of the Conservation Voters New Mexico Education Fund, sister organization to the nonprofit Conservation Voters New Mexico. At the end of September 2015, Juntos celebrated the pope's visit to the United States by projecting his address to the joint session of Congress on the wall of a long, narrow office above a pizza joint in Albuquerque. Spanish and English whispers filled the room as Pope Francis began to speak, citing four import-ant Americans: Abraham Lincoln, Dr. Martin Luther King Jr., Catholic Worker Movement founder Dorothy Day, and the Trap-pist monk Thomas Merton. An older woman sat, rapt, while the young woman who accompanied her pulled out a phone and typed, "Who is Thomas . . ."

In "On Care for Our Common Home, Laudato Si," Pope Francis wrote again and again that ecological action and social action must be one, and that the most vulnerable people must be protected. "Intergenerational solidarity" comes up too: "Leaving an inhabitable planet to future generations is, first and foremost, up to us," he wrote. "The issue is one which dramatically affects us, for it has to do with the ultimate meaning of our earthly sojourn." The refusal to reverse the planet's warming trend and the refusal to eliminate poverty are related by the lack of political will and also by the fact that some profit while many suffer. Imbalances between the global north, with its capital and military might, and the global south, with its vast natural resources, perpetuate an "ecological debt." And, he wrote, blaming population growth rather than "extreme consumerism" for problems is a refusal to face the real problems. "It is an attempt to legitimize the present model of distribution, where a minority believes that it has the right to consume in a way which can never be universalized, since the planet could not even contain the waste products of such consumption."

After the end of the pope's address to Congress, Patricia Gallegos, lead organizer for Juntos, came to the front of the room. No longer a practicing Catholic herself, Gallegos told the group it's nice to listen to the pope's words—to hear what he is saying about extractive industries and the economy and how it affects poor people, people of color, and the earth itself. But it's crucial to transcend words and take action, she said.

At the viewings and the weekly study-group sessions, people seem to feel a mix of excitement, inclusion, and hope. The pope's message of social justice appeals to the young people. Older environmental activists are drawn to the discussions of sustainability and ecological awareness. And people of faith are listening to the messenger.

Quintana pointed out that the message within the encyclical is serious. But it isn't one of despair. "Because it's easy to feel crestfallen, he calls us to hope and joy," Quintana said. "After all, as

Pope Francis wrote midway through the letter, 'Injustice is not invincible.'"

———

THE POPE'S ENCYCLICAL letter didn't shed new light on the changes taking place across the planet, but it is framed in a new way, said Joan Brown, executive director of New Mexico Interfaith Power and Light, a nonprofit that advocates for energy conservation, efficiency, and renewable energy. "He's looking deeply at a whole new way for human beings to live, and at the structural issues intertwined with ecological issues," she said. "It's very poetic and written very beautifully, as he is looking at the good of the community and the joyfulness and hope amid the difficulty."

Brown grew up on a family farm in Kansas, at the edge of the bluestem prairie. Her Catholic faith has always guided her, and the sixty-one-year-old Franciscan nun is a common sight at public hearings and protests. "My spirituality is one that is rooted in love of creation, of humanity, of justice," she said, looking out at the small garden at the back of her home in Albuquerque. "What keeps me going is prayer, meditation, working for justice, and hands-on caring for the earth."

To her, Pope Francis defines the entire encyclical in the first paragraph, when he refers to the "Canticle of the Creatures," a song from the early thirteenth century, writing, "In the words of this beautiful canticle, Saint Francis of Assisi reminds us that our common home is like a sister with whom we share our life and a beautiful mother who opens her arms to embrace us. 'Praise be to you, my Lord, through our Sister, Mother Earth, who sustains and governs us, and who produces various fruit with coloured flowers and herbs.'"

"'Sustains and governs us'!" Brown's face broke into a smile, and she leaned forward in her armchair. She has a way of being warm, calm, and excited all at once. "We usually don't think of the earth as governing us. It kind of startles me every time I read it."

When Santa Fe's patron saint wrote his canticle—or chant—to which Pope Francis refers, his health was failing, and regional conflicts were simmering. Symbolically, to Brown, it was a time that is similar to this moment in history. Right now feels like a dark time. A time of crisis. Yet if we have grateful hearts, she believes great wonder, beauty, and healing can come from the darkness.

Brown noted that despite the pope's encyclical and its focus on the earth, climate change, and social justice, the US Conference of Bishops, which sets the tenor for Catholic churches in the United States, chose to stick to its own agenda—emphasizing issues like reproduction, pornography, and opposition to gay marriage. But more and more often, local clergy and congregations are talking about climate change, she said, and many are taking on energy-efficiency projects, installing solar systems, or working on congregation-wide conservation programs. "It's a mixed bag: Some are seeing it as an opportunity to live out our faith more, and others are not embracing it," she said. "But we need to connect the dots on these issues, and I think we need to use climate change as an organizing principle."

The immigration and refugee crises will only intensify as people living in arid lands can no longer survive or as the seas rise, inundating islands and coastlines, she said—a fact that's borne out in the years that follow our conversation in 2015. That sort of destabilization can lead to increasing violence, human trafficking, food insecurity, and growing poverty. "With our brothers and sisters suffering across the planet, religious people—who already often work with food aid, education—need to make the links," and she added that making those connections is especially important here in New Mexico. "In this state, it's always the political or economic interest that trumps the deeper values we have, values for the common good."

That can apply to everything from opposition to the EPA's Obama-era rules to reduce greenhouse emissions from industry to the state's lack of water planning. For decades New Mexico has been a sacrificial zone, she said, ticking off just some of the problems: more than a thousand uranium mines that have never been cleaned up, land and water pollution across the state, coal-fired power plants and the methane hotspot in the Four Corners, the Waste Isolation Pilot Plant, and widespread oil and gas drilling in southeastern and northwestern New Mexico. "We're so used to looking at things piecemeal that we don't look at the whole," which is what the pope is asking us to do, she believes. "We need to start addressing this from the heart. We have the facts. Now, we need to think of love as a call for ecological action."

Looking toward the future of the state's economy is a part of that. "I continue to feel so badly: We are always last, or close to last, in the nation on poverty, all these things," she said. When elected officials and others say that the energy or mining industries need special support—or regulatory shortcuts—because they provide crucial funding for the state's budget, Brown doubts the wisdom of that. "I keep thinking: We've been doing this for decades— and we're still last."

In the spring of 2015, Interfaith Power and Light wrote to each of the eighty elected leaders in New Mexico who identify themselves as Catholic. Signed by more than one hundred of the state's faith leaders from churches, synagogues, and meeting houses, the letter invited them to heed the pope's call to action on climate change. "No one responded," Brown said, though she added in her gentle way, "I'm sure that they are busy." Meanwhile, she thinks solidarity is growing among New Mexicans. People of faith and people who care about the earth (or who are in both camps) are coming together, especially as non-Catholics embrace Pope Francis's message on climate change and social justice. She's hopeful that at the upcoming United Nations talks in Paris, countries of the world will come together and act on climate. "It's an exciting time to be alive," she said. "There's a lot of grace in it."

———

ON A CHILLY night at the end of November 2015, dozens of people met at the corner of Sixth Street and Gold Avenue in downtown

Albuquerque. A few young children delighted in holding flickering, battery-powered tea candles, while adults illuminated the pamphlets that have been handed out with their cellphones. People had gathered to pray for the United Nations climate-change meeting that was about to begin in Paris. After a few words and songs—and a look around to see if anyone else was on the way—they walked in solidarity with climate refugees and immigrants to Immaculate Conception Catholic Church a few blocks away.

Inside the church, more than a hundred people shuffled into the front pews. There were older men sitting alone, mothers cupping the heads of their daughters as they whispered to one another, families with teenagers, couples, and small packs of friends. Candles were lit. Some people bowed their heads.

Temple Bnai Israel's Rabbi Arthur Flicker issued the call to prayer by blowing the shofar, or ram's horn. First, a Buddhist prayer is offered. Then, Ruby Kochhar sings a haunting Sikh chant, and Rev. Sylvia Miller leads a sacred dance. Necip Orhan, with the Dialogue Institute of the Southwest, recites a Muslim prayer. Later, he explained that he likes what the pope is saying and has written about climate change. "Today," Orhan said, "all Christians and Muslims are sisters and brothers. The encyclical, it is exactly what we want to say, what we want to do."

The tone of the evening, and the mix of people within the church, reminded me of how in the days following September 11, 2001, people of many faiths came together to pray, mourn, sing, cry, and hold hands. Even in my small hometown in Connecticut, where I awaited a rescheduled flight out of Hartford while visiting my parents, people of diverse faiths crowded into the Congregational Church on the evening of September 12. Looking back on that night, I remember two things most clearly: Holding my father's hand. And how tender people were with one another. Something terrible had happened. And we wanted the comfort of one another's faith and presence. Similarly, the mood in 2015 is somber. This is not a celebration; nor is it a rowdy protest.

Toward the end of the interfaith service in Albuquerque, Father

Warren Broussard, pastor of Immaculate Conception, blessed Brown, who would leave in the morning for Paris and the climate talks. He asked everyone to raise a hand as he recited the blessing. As he intoned the words, some people tired and lowered their arms; others held their lit candles in one hand, keeping the other raised toward Brown. All eyes were upon her. At the end she pressed her fingers beneath her glasses, wiping away the tears.

When musicians returned to the front of the church, the two men led a closing song: "This is my song, O God of all the nations / a song of peace for lands afar and mine."

Many didn't know it but followed along with the words printed upon the paper. And even those unsure of the tune raised their voices together in song.

————

I'D LOVE TO end the story there. With Brown illuminated in light on the altar, lifted by voices in song. Or with the headlines following the Paris meetings, when news organizations pegged the Paris Agreement as "historic," and the head of the World Bank called it a "game changer."

The reality, of course, is a bit muddier.

"The agreement is a clear articulation of the fact that almost all countries of the world now agree that climate change is a serious problem—and more serious than they agreed to in the past," climate scientist Jonathan Overpeck told me over the phone in 2015, right after the meetings ended.

By recognizing the magnitude and importance of climate change, he said that delegates in Paris did achieve a "major milestone in human history." But, he added, "the devil is in the implementation." Countries sent their delegates to Paris with pledges and plans to cut greenhouse-gas emissions, but the agreement lacked any enforcement mechanisms. Pledges were voluntary, and there were no penalties for not reducing emissions.

Delegates recognized that trying to restrict warming to 2°C will not protect many countries, especially island and coastal nations that are being inundated by sea-level rises. Instead, they said it was imperative to limit global warming to 1.5°C above pre-Industrial levels. That's going to be pretty tricky, not only because the agreement lacks binding emissions reductions, but because the voluntary pledges still won't reduce emissions enough to limit warming to even 2°C.

In addition, the United Kingdom's Met Office announced at the same time that Earth's global average temperature is poised to reach 1°C above pre-Industrial levels. "The first thing to realize is that this year [2015] is going to be a real eye-opener," said Overpeck. "We're going to see the world-temperature record broken, or more like shattered. It's going to be in that 1° range, if not by the end of this year, then very soon thereafter. That, from a scientist's point of view, is pretty scary."

The question is, then: Can humans really limit the temperature rise to 2°C, never mind 1.5°C?

In 2015 Overpeck said most scientists think that's still achievable.

"There might have to be an overshoot: It might have to get warmer than 1.5, and then it comes down pretty quickly," he said. "If we're able to do that, we may be able to preclude massive melting of the ice sheets and other effects."

The technology exists, he said. The question is whether there's political will, on the part of the United States and nations across the world.

Although the number of Americans denying or ignoring climate change has become increasingly small, they still have a great deal of influence on policy, which affects people across the globe, he said. "That ranges to people on islands in the Pacific who will be flooded, who will lose their whole country, to people in semiarid regions, where 'hot drought' could lead to all sorts of societal stress and mass migration," he said.

Then, in June 2017, of course, Trump pulled the United States out of Paris altogether.

FIGURE 7. Jemez Mountains. Photograph by the author.

6. NO WAY BACK

FIRST THERE'S A spark, and then the fire. We all stare at the sky, smell the smoke. After the trees and brush and roots are gone, floods roar through arroyos and down hillsides. Weeds invade as soon as the ground has cooled.

Often, the long-term changes aren't that obvious, especially when compared to flames and floods. But what's been happening across tens of thousands of acres within the Jemez Mountains isn't subtle. Nor are changes happening slowly.

In what amounts to the blink of an eye, the Jemez Mountains have experienced landscape-level changes in their forests and watersheds. Some of the forests New Mexicans have known for generations won't ever return. "It's easy to look out here and see all the dead trees and feel all bad about it, all depressed about it," said Collin Haffey, at the time an ecologist with the US Geological Survey. "It's harder to see the remnant forests."

In April 2017 Haffey brought a team of artists and conservationists to those remnant forests, indulging a journalist in tow. On the day before Easter 2017, we visited the "heart of darkness"— that's the name Haffey and some of his colleagues have given a 33,000-acre area of the Jemez Mountains scorched by an inferno in the 2011 Las Conchas fire.

SINCE THE 1980S an increasing number of big fires of over a thousand acres have been burning in the western United States. They come on the heels of decades of fire suppression in the forests. And as the climate warms, the sheer number of fires has grown, too, and the wildfire season has lengthened by about two months. Even in years when there is a bumper snowpack in the mountains, the National Interagency Fire Center still often predicts significant wildfire danger for southwestern and central New Mexico.

Flanking Santa Fe, the Sangre de Cristos Mountains have experienced big blazes in recent years. And the 2012 Whitewater-Baldy Fire burned nearly three hundred thousand acres of the Gila National Forest. But perhaps no place better illustrates how southwestern forests will continue changing in the warming world than the Jemez Mountains.

The Jemez forests have been hammered by bigger and nastier fires since the late 1990s. To mention just a few: In 1996 the Dome Fire burned about 16,000 acres of the Santa Fe National Forest and Bandelier National Monument. Started as a prescribed fire, the 2000 Cerro Grande burned out of control, lighting up 48,000 acres, destroying or damaging 280 homes in Los Alamos, and wiping out 40 buildings at the national laboratory. Then in 2011 Las Conchas devoured 156,000 acres, with flames visible from the Santa Fe city limits. It's worth repeating that the fire's behavior, growing from 40 to 43,000 acres in its first fourteen hours, defied the rules. That fire also changed our expectations when it comes to fire—and what comes back after the flames.

Fighting disasters like Las Conchas is expensive. Controlling and extinguishing that one fire cost almost $50 million; emergency rehabilitation on the burned areas cost another $2.5 million. The fire destroyed dozens of homes, and the floods that followed the fire destroyed more buildings as well as privately owned orchards in Dixon and the Pueblo of Santa Clara's watershed.

The total costs of the fire and its aftermath haven't been calculated in full. But the cost of fighting wildfires in general is rising, in part because more people live along the fringes of the forests, forcing firefighters to defend properties and communities when those bigger, more frequent fires occur due to warming. Nationally, more than 10 percent of the US population lives within these zones—and developments are increasing all the time, despite the risks from wildfires.

———

FROM HIGHWAY 4 in the Jemez Mountains, we turned down a gravel road where live ponderosa pines still tower above, blocking out the sky. Las Conchas burned through here, but it was still cool and green, shady and quiet. Haffey pointed out a meadow where high school kids come to party. It was green, dotted with the spring's first dandelions. As we continued southeast, the vista opened up. Nothing stood taller than a basketball hoop, except for spindly, blackened tree trunks. The skeletons with thick trunks and black branches are ponderosas, Haffey pointed out. Douglas firs have skinnier trunks and white branches.

On one side of the road, aspen trees were beginning to spread light-green leaves to the spring. On the other, the predominant plant was the thorny locust. When Haffey was out here ten days after Las Conchas burned through, the locust sprouts were already a foot tall. "That was with zero rain," he said.

But there were also rabbits and squirrels here, and a chipmunk ran past, its tail poking straight into the air. Where the aspens were taking over, it felt like this would be a forest someday. Not a ponderosa forest, but a forest nonetheless.

Kathleen Brennan, a documentarian, and Shawn Skabelund, an installation artist, are on the trip as part of the East Jemez Landscapes Futures project, which Haffey hopes can incorporate art and storytelling into land management. He was taking them into the burn scar so the artists can see what's been happening on the ground since Las Conchas and the Dome Fire. "Art and story can be used in ways to inform, engage, and involve a community, and in Northern New Mexico, people have really strong ties to place,"

Haffey said. "I want art to be that conduit between communities, scientists, and managers, and back and forth."

He envisions an art collaborative that solicits art from local communities that can be exhibited in lobbies and visitors centers. "That would allow for people, as individuals or groups, to tell their story of how those big fires and landscape-scale changes affected them." Fires like Las Conchas are such a massive disturbance, he said—not just the fire itself, but its aftermath, too—that encouraging communities to tell their various stories can give them scale. This idea isn't separate, in his mind, from managing the land for healthy fire, species diversity, and a warmer future.

Art and storytelling could be fundamental to the management process, too, not separate from it. The most famous land manager, he points out, is Smokey Bear. And he wonders what the new story is, the one that will continue the story of the landscape in a way that's historically significant and that represents the transformation of the landscape and its communities. "We've all been to workshops with managers and scientists," he said, and he added that maybe some of those scientists should be replaced with artists who will generate different conversations. He thinks that's one way to move forward as a community, as scientists, and as land managers—to integrate not just science, but also art and story into land-management strategies.

Driving further, we crested a hill. And everything changed.

Six years after Las Conchas, the hills are still stripped bare. Looking out across tens of thousands of acres, every canyon, divot, and lump on the landscape is visible. It's like looking at an interactive map. Slide the cursor to the left to see the vegetation; slide it right to see the geology. Only this is real life. What used to be a ponderosa forest now looks like the Guadalupe Mountains in West Texas.

"I've probably taken about a hundred people out here," Haffey said. "I never get used to the scale of this." He points to the canyon below, where we can see the matchstick skeletons of ponderosas. "Those survived the Dome Fire. Hand crews saved those stands."

But then Las Conchas roared through.

Haffey talked about those wild initial days of the fire, which broke out at the end of June. "It was still burning at midnight, long after normal fire weather," he said. "Las Conchas broke all the rules. There were huge plumes blowing up, and then there'd be nothing to burn and the columns would collapse. It was like fluid dynamics more than fire." Burning gases shot fireballs hundreds of feet into the air.

He described seeing needles still on the trees right after the fire. They were flash-frozen, a grayish color. When the fire roared through the trees, it burned the windward side and created a back eddy on the leeward side, baking the needles rather than burning them.

Then we descended into the heart of darkness. We passed the spot where an abandoned campfire above Cochiti Canyon ignited the Dome Fire, then we parked again. Everyone took their time looking across the canyons, scuffling through the sands, and pointing out scat or flakes of obsidian. In the canyon below, there were a handful of bighorn sheep.

The species had been extirpated from the area in the early twentieth century. Biologists talked about relocating a population here earlier, but thanks to fire suppression and the dense forest that had grown up through the canyons, it wasn't a suitable habitat for the sheep. After Las Conchas, however, the New Mexico Department of Game and Fish moved Rocky Mountain bighorn sheep from the Wheeler Peak Wilderness to Cochiti Canyon.

Six years ago we wouldn't have been able to see more than a dozen feet ahead of us. Now we looked down the canyon, and, hundreds of feet below, the sheep peered up at us.

Then we took off across a wide plain and up to St. Peter's Dome. That April day, the dome was a dusty knob on the eastern edge of the Jemez Mountains. In the Rio Grande Valley below, Cochiti Reservoir was visible. One side of the dome used to be covered with ponderosas and pines. On its other flank would have been a mixed alligator juniper and piñon forest.

Now, after the Dome and Las Conchas Fires, patchy grasses

push through the tan soil, and scrubby oaks rattle with last fall's orangish-brown leaves. "Historically, this was a frequent fire area," Haffey explained, but it still would have been a pine forest. He didn't live here before the Dome Fire, but from what he's heard, this would have been a darker, more humid spot than it is today, when we're gritty from the wind and dust. "We would be standing in the shadows, it would be hard to see through, crawl through."

The biggest change since the Dome Fire, he said, was to those slopes miles to the west. "That was ponderosa as far as you can see."

Haffey thinks it's pointless to feel depressed. It was the first day of turkey season, and in addition to the hunters we saw, three trucks were parked on the dome, about ten miles from the paved road, at the trailhead of a hike into Bandelier. While we ate lunch two more guys pulled up, hoping to get through the locked gate and four-wheel to the top of the dome. People talked about the weather or the view or asked for directions. Not one person we met mentioned the landscape or the fire scar.

There's no returning the forest to what it once was, Haffey said. But the area can still be managed in a way that promotes a diversity of species. He and Sasha Stortz, from Northern Arizona University's Landscape Conservation Initiative, talked about how some places out here still haven't been surveyed or studied since Las Conchas and the floods and debris flows that wiped out fish populations. No one knows what might have come back, since only bacteria and viruses likely survived.

"People say, 'You don't know what you've got until it's gone,'" Stortz said, "but you also don't know what you have until you know what was there before."

Haffey nodded and added, "And you don't know until you go and look."

———

SOMEONE WHO HAS gone back, way back, is USGS research

ecologist Ellis Margolis. He wanted to know not just where fires have burned, but how they burned. He also wanted to know if Las Conchas really was an extraordinary fire for the Jemez, or if fires burned like that before suppression and before they were recorded in government reports and news stories.

"People say, 'The trees will come back. They always have,'" Margolis said when I met him at his office in Santa Fe. "We're digging in, to see." There are multiple layers of evidence, including historical photos and written accounts, tree rings and fire scars and paleo-charcoal sediment records. "When you smash all those together," he said, "you can get a really good record of what's going on in any one location."

We've all pressed our fingers to tree rings, looking for the fat rings representing wet years and the slim lines showing when times were dry. Some trees also respond to summer and winter precipitation, explained Margolis. Looking at those samples, dendrochronologists can gauge snowpack and the strength of the monsoons.

By cross-dating tree-ring samples, Margolis can tell the story of fire in the Jemez from the present day all the way back to the mid-1500s. He and his colleagues have even identified some samples from the 1100s. In that record, he identified more than one hundred fires. Most of those were low-severity fires, in ponderosa and mixed conifer forests.

The year 1664 was his favorite, he said. "It was burning through the whole year. The monsoon was weak or shut down that year." Another in 1729 had a huge footprint, likely bigger than the size of Las Conchas. And in 1752 a midsummer fire burned just as big as Las Conchas, too, but it was low severity. In the following decades there were more; their scars fit together like a jigsaw puzzle on his map. Following a wet decade in the 1790s, a huge fire in 1801 burned about three hundred thousand acres. "There was fire all over the place," he said. "That should be our goal, to get the good fire back in."

So what about Las Conchas? Was it an uncharacteristic fire for the Jemez?

In the record it's not unusual to find fires bigger than one hundred thousand acres. But they were low severity. Big fires burned in regular cycles, but they didn't change the makeup of the forests.

"Once you wipe out the seeds' source," he said, "in a hotter environment, you can't get those trees back into those areas, especially as it gets warmer and drier."

In other words: No, those forests aren't coming back.

AFTER SAYING GOOD-BYE to Haffey, Stortz, and the other scientists and artists, I returned down the gravel road, past that patch of ponderosa pine that survived Las Conchas. Water bubbled up from the ground where the last of the snow patches were melting or little seeps emerged from the ground. The air smelled like vanilla, and as the wind blew it brushed through the green needles with a whisper.

I drove past the meadow and through the thickets of oak and locust. At the turnoff to a road near Capulin Canyon, I parked and walked. There was a high whistle behind me, but even when I stopped and turned, I couldn't figure out where it was coming from. It disappeared by the time I crossed an open patch of grass and came to a stand of a dozen live ponderosas. They were maybe a hundred feet tall, their trunks mostly bare of live branches. But their crowns were tufted with live green needles.

Everywhere else there were waves of oak and walls of locust, punctured by the occasional charcoal-black trunks of ponderosas poking up like tapered, burned-out candles. Some looked like sewing needles, blue sky peeking through holes eroded through the tops. One trunk rose fifty or sixty feet. Most of it was scorched, but a little more than halfway up from the ground there was a chunk of brown wood still wrapped with bark. It was like a marshmallow shoved too far down a roasting stick.

From the edge of Alamo Canyon, White Rock was visible over

the lip of a ridge, and the Sangre de Cristos still had snow on their peaks. Across the entire vista, downed ponderosa trunks pointed downhill on both sides of the canyon. The town below was spotted with white buildings, and the mountains in the distance were blue and white. But everything else was brown or gray—until I spotted, on the near flank of the canyon below, the light-green fuzz of aspen saplings.

In the roadcut there's coyote scat, and above, five or six turkey vultures circle, an ambassador occasionally breaking ranks to check me out. A pair of robins keep watch, and one nosy little bird—too far away to identify even with the camera's zoom lens— follows, flitting from the top of one scorched snag to the next.

There's also a red-tailed hawk and what looks like the peregrine we saw earlier in the day below St. Peter's Dome.

Back at the truck I realized my skin was coated with sand, and I heard that whistling sound again. It's the wind from Capulin Canyon passing through the ponderosa skeletons.

This isn't a pine forest anymore.

Tens of thousands of acres of forests are gone from the Jemez Mountains. In our warming world, they won't grow back, and the evidence of that was clear today.

Leaving the burn scar and heading toward Jemez Springs on Highway 4, I didn't relax into the landscape the way I thought I would. The forest looked too dark, too dense. I couldn't imagine living within this thicket. Where the Jemez River chugged alongside the road and the canyon widened out, the slopes above the road were monotonously green, crowded with tall pines.

These probably aren't the forests of our futures. And I finally, viscerally, understand what firefighters and land managers have been saying for decades: This forest is not sustainable in our warming world. It took just one day in the burn scar to understand that.

———

A FEW MONTHS later Haffey and I met again at Bandelier National Monument. Known for its archaeological sites, including dramatic cliff dwellings above Frijoles Canyon, the monument covers thirty-three thousand acres in the eastern part of the Jemez. Las Conchas burned more than 60 percent of its area.

The fire was bad enough. Then came the floods.

In 2013, two years after Las Conchas burned through about three-quarters of Frijoles Canyon, wild floods roared down Frijoles Creek, ripping up cottonwood trees and dragging giant ponderosa pines from miles upstream. When we visited four years later, the stream bed still had piles of brush and twisted tree limbs—some taller than a house—that showed where flood water ripped through. The scale of the destruction is enough to make even a tourist to the archaeological sites nod and think, *Ah, that's why people built their homes and villages so high above the tiny stream.*

Bandelier's chief of resource management, Jeremy Sweat, talked about the deep emotional connection people have to Bandelier and its landscape of rugged canyons and mesas. "You talk to people in the community, and they tell stories about fishing with their grandfather and the beaver ponds up on the Frijoles. They talk about hikes they took as children to go to the waterfalls here at Bandelier or into the backcountry as Boy Scouts," he said. "Those are really important connections people have, and within less than a generation these places have changed so drastically."

Areas that used to be ponderosa pine or mixed conifer forests have now transitioned into grasslands and shrubs. "We don't know if those forests are going to regenerate naturally, we don't know if those shrublands are going to be sort of the more permanent type of forest that we have," he said. "We're really trying to figure that out right now."

Sweat also pointed out that the changes aren't unique to Bandelier and the Jemez Mountains, or even to northern New Mexico. "These types of changes, with high-severity wildfire; longer, hotter droughts; and impacts to vegetation are happening all over New Mexico and all over the American Southwest," he said. As temperatures keep rising and droughts become more severe, there will be more fires and more changes all over the western United States. "I think it's important for all of us to care about these changes and what they mean, not just to recreation but to local communities who have these emotional connections to the place, and the tribes who have used these lands and collected plants and materials here for hundreds or thousands of generations."

Las Conchas also burned 50 percent of the Pueblo of Santa Clara's watershed, according to Matt Tafoya, the pueblo's forestry director. And together, Las Conchas in 2011, Cerro Grande in 2000, and the Oso Complex Fire in 1998 burned 80 percent of the tribe's forests and about 80 percent of the tributaries to Santa Clara Creek, which flows to the Rio Grande. "This is our ancestral homeland," I heard Tafoya say during a public meeting. "This is a resource used on a day-to-day basis by members of our community."

Again, people weathered the fires, only to face floods. With trees and grasses gone, water ripped down arroyos, moving in dangerous, roiling sheets down the mountainsides. During his presentation at the meeting, Tafoya played a video he filmed while waiting for one of those floods after a summer storm delivered about a half-inch of rain upstream. The clip opens with a dry arroyo and pine trees in the top third of the frame. "Watch the trees," Tafoya said. Suddenly, churning black waters fill the top of the frame. The video flickers off and then changes perspective as Tafoya gets into his truck. Within seconds the water, ash, and debris has roared across the road. The camera remains steady while the truck backs away from the water heading toward it on the road. Tafoya filmed that video more than two years after Las Conchas burned through the watershed.

Before Las Conchas, the largest flood events for Santa Clara Creek were predicted at about 5,600 cubic feet per second (cfs),

Tafoya said. Now, that predicted "hundred-year flood event" is estimated at more than 21,000 cfs. Meanwhile, where it flows through the village, the river channel has the capacity hold about 9,000 cfs. Set aside those numbers for a moment, and think about it this way: For centuries, the channel could handle twice whatever its biggest flood might have been. Now that's no longer the case.

———

ON THAT VISIT back to the Jemez in July 2017, I caught a ride with Haffey from Bandelier, and we retraced our route from Highway 4 onto the Dome Road. Driving through the Dome and Las Conchas burn scars, we talked along the way about climate change. At thirty-seven, Haffey has really only known the Jemez as a place that's experiencing rapid change. There's no way to bring back the ponderosa and mixed conifer forests in parts of the Jemez. Dropping seeds from airplanes or engaging in massive replanting efforts would be expensive—and it's not likely anyone has that kind of money to spend, especially the perpetually cash-strapped Forest Service, which has more and more wildfires to worry about each season.

It's not even clear that money would be well spent. "We don't know if this landscape can support ponderosa pine in the future," Haffey said. "Planting trees now that will die in a drought in fifty years may not make the most sense." Maybe it's better to look around and understand what this new system means and provides, and how these new systems might make it through droughts. "Maybe these shrublands and grasslands are actually a much more resilient system in the future," he said.

When we pulled over above Cochiti Canyon, I wanted to see the ponderosa pine Haffey had pointed out earlier in the spring. It's the oldest live tree out here in the heart of darkness—a five-year-old ponderosa pine that reseeded naturally after Las

Conchas. I remembered it was near elk scat and a flake of shiny obsidian. I remembered that when I stopped to snap a picture of Haffey, the other biologists, and the two artists with the tree, a hummingbird buzzed us. But this time I couldn't find it.

Then Haffey pointed to the tree. It was taller than I'd remembered, and I'd been looking too low.

FIGURE 8. Tom Swetnam's tree-ring laboratory. Photograph by the author.

7. RED FLAG WINTER

ON THE LAST day of January 2018, Kerry Jones pointed to signs near the Crest Trail in the Sandia Mountains. Nailed to a conifer, the signs guide cross-country skiers along the trail, and they're placed high enough to be visible in the winter snows. On this day, the signs were a good three feet above his outstretched arm. Beneath his boots, there were only a couple of inches of snow.

It's not supposed to be this way.

A few hundred yards away at Sandia Peak, the view was even more grim.

Thousands of feet below, the Rio Grande Valley looked dusty and dry. To the west Mount Taylor should be a hulking white mass this time of year. Instead it was just a deeper shade of blue than the sky. Along Sandia Crest, what snow there might have been has blown back from the edge. "We're up at just a little bit above ten thousand feet, and in the world of weather this is high altitude," said Jones, a warning-coordination meteorologist with the National Weather Service in Albuquerque. "This time of year we should be not on bare ground as we are today, but [standing in] several feet of snow."

The lack of snow in New Mexico's mountains would have implications for farmers and cities in the spring and summer. Certain tree populations in many of the state's mountain ranges, including the Sandia and Jemez Mountains, are already experiencing large-scale die-offs. And communities should be preparing for wildfire season. "We are standing at the driest start to any water year on record in the observational period, which goes back to the late 1890s," Jones said. "There is no one alive today that's seen it drier for any start to a water year."

A water year begins October 1 and helps meteorologists, water managers, tribes, and various agencies plan. Jones explained that New Mexico's snow-accumulation season typically begins in late October. From there, snowpack is supposed to build through early spring. Jones said storms could still hit the Sandias and other mountains in New Mexico this spring—and recent storms brought one to six inches to places like Gallup, Santa Fe, and Angel Fire. But new snow can't make up for the existing deficit. "We would basically need two and a half times our normal precipitation for northern New Mexico into southern Colorado to even bring us back to where we should be this time of year," he said. "When you think about it, that's just a tremendous deficit to overcome." It would be "pretty unprecedented," he said, to get that much snow between now and early April.

On that January day, the National Weather Service had even issued a Red Flag Warning for most of eastern New Mexico. These warnings alert people to critical fire weather patterns and usually start in the spring. It was the sixth one they issued in January. Winds increase fire danger and whip up dust storms. They also dry out soils and whisk snows away before they can melt and make their way into streams and rivers. That's a problem as New Mexico's springs continue to become warmer earlier. "As we've seen for the last several years, we're not getting as much snow—and it's warmer, and so we're melting that snow much earlier," said Jones. Instead of snowmelt peaking in May and June, when farmers need that water for their fields and orchards, the waters are churning in late February or March.

People are worried about fire season, too. In 2018 it began in March—rather than May or June, as in the historic past.

"This is a pretty scary year," said Tom Swetnam, regents'

professor emeritus of dendrochronology at the University of Arizona, where he directed the Laboratory of Tree-Ring Research. "Typically, the hottest fires occur after the driest winters and driest springs, so the signs are not good. We're very concerned right now because the moisture is very low, and most of the weather stations around the state are showing that we're in one of the driest—if not the driest—winters on record."

The connection between drought and fire is obvious in the twentieth and early twenty-first centuries—just think about Las Conchas. The connection is also visible within the tree-ring record, which Swetnam and his colleagues have been studying for more than thirty years. And their records go back centuries. "A lack of moisture in the cool season preceding fire season is really key," said Swetnam. Dry winters and springs are a recipe for bigger, hotter fires in the Southwest.

Overlaid onto cycles of drought is long-term warming. As the earth continues warming, droughts aren't just dry. They're also warmer than they used to be in the past, Swetnam said. "Now, when we flip into a drought mode on a yearly or seasonal basis, on top of that, the magnitude of the drying is exacerbated by warming," he said. "It makes it even more intense."

Tree-ring scientists, like Ellis Margolis with the US Geological Survey, peer into the past to understand the impact of drought on forests. But Swetnam can also look outside his window to see how forests in the Jemez Mountains have changed over the past few decades. Swetnam grew up in Jemez Springs and moved back here from Tucson a few years ago. His father was a forest ranger on the Jemez Ranger District, and he recalls playing in the snow on the hillsides off the highway between Jemez Springs and the Valles Caldera. Those same grassy valley bottoms and rounded hills are now covered in trees. "There's been an increase in density just within my lifetime," he says. "It's clearly noticeable."

After decades of fire suppression and overgrazing, the forests became far too dense—and those fuels also increase the risk of high-severity fires.

Swetnam warned against overgeneralizing across all forest types and elevations in the Southwest. But the Jemez is the "poster child for high-severity wildfire" in New Mexico. "One after another, fires have been eating away at the forest up here, creating giant canopy holes and destroying hundreds of homes," he said, "with the potential of losing thousands of homes." From his deck he points to where Redondo Peak sports a scar from the 2013 Thompson Ridge Fire, which burned about twenty-four thousand acres of Valles Caldera National Preserve and the Santa Fe National Forest. And he recalls the 2000 Cerro Grande Fire that destroyed hundreds of homes in and around Los Alamos. "That's what we're facing as the temperatures keep going up and we're getting drought years like this one," he said. "We're going to lose more neighborhoods in the Jemez . . . and I hope we don't lose human lives."

Down below his deck the Jemez River cuts a course through the valley and roofs poke out from the dense forest. Swetnam estimated there are more than a thousand homes built within overgrown ponderosa pine forest in the Jemez. Some of these neighborhoods can only be reached by steep, dead-end roads. That means there's only one way in—and only one way out. Severe, quick-moving fires can trap people, including firefighters, in those neighborhoods.

Each year as fuels build up, drought intensifies, and temperatures keep warming, the risk for severe fires grows greater and greater. This isn't unique to the Jemez. Wildfires across the West have grown in number and size since the 1990s—a trend that's predicted to continue as warming continues.

The US Forest Service and its partners have been working to thin the forests. Swetnam is encouraged by programs like the Southwest Jemez Mountains Resilient Landscapes and Collaborative Forest Landscape Restoration Project. That ten-year project was borne out of the 2009 Forest Landscape Restoration Act, championed by then-senator Jeff Bingaman. That act promoted collaborative, science-based ecosystem restoration in certain areas identified by the Secretary of the US Department of Agriculture, which oversees the Forest Service, as priorities across the nation. The Southwest Jemez project is one of twenty-three such projects nationwide.

The project covers 210,000 acres spread across the Santa Fe National Forest, the Valles Caldera, and the Pueblo of Jemez. Since natural disasters like wildfires as well as floods and insect infestations don't stay within fence lines, the Southwest Jemez Mountains Resilient Landscapes and Collaborative Forest Landscape Restoration Project includes about three dozen federal, tribal, state, industry, nonprofit, and university partners. Their goal is to bring the Jemez's forests closer to what the forests were like before industrial-scale grazing in the late 1800s and a century's worth of fire suppression led to the thickets they had become by the end of the twentieth century.

"There are other organizations, like the Nature Conservancy and others, that are working hard to try to turn things around in terms of the resiliency of our forests, and make them more resilient," Swetnam said. But people living within and alongside the forest need to take fire seriously.

"We all have a collective responsibility to deal with the problems, especially if you have a home in the forest—it is your responsibility to reduce the risk to your home, your neighbor's home— and the firefighters who would have to come in and try to save houses," he said.

To protect homes, communities, and firefighters from moderate-severity fires, Swetnam said people can thin trees and make their homes "firewise."

"A firewise home means that it's less likely to catch on fire as a fire approaches," said Swetnam, who has been helping raise awareness around his own community. Basic steps include moving exposed wood and firewood at least fifty feet from structures, using metal or asphalt shingles, and burying propane tanks. "If fire gets under those and starts heating them up, they will pop off a safety valve and start shooting flames out of the top of them,"

Swetnam said. "Or, if that safety valve doesn't work like it's supposed to, it'll blow up and it'll go off like a bomb."

Swetnam and his wife have thinned their property, leaving only the strongest and oldest ponderosa pines and clearing out the "doghair thicket" of stunted and spindly young trees.

Before widespread livestock grazing, and then the late nineteenth-century Forest Service policy of fire suppression, ponderosa pine forests looked more like this: older, widely spaced trees with grasses and forbs running beneath them. Ponderosa forests thrive when periodic, lower-intensity fires burn through the understory. Based on the tree-ring record for the Jemez, those types of fires typically burned through areas once or twice a decade, Swetnam said. As fuels built up, however, fires changed, roaring into the canopies and destroying entire forests.

Thinning doesn't mean clear-cutting. "Most of the trees can stay," said Swetnam. Thinning can cost anywhere from five hundred to a few thousand dollars an acre, he said, adding that many local contractors will come out for an evaluation and free estimate. There are also state or federal grants to help with the costs, and homeowners may get lower insurance rates if they take certain steps to protect their homes ahead of wildfires. "The insurance companies are starting to take notice, especially of fire risk, and in New Mexico we're starting to see rates are going up," he said. "Some of this is, I think, because of the loss of homes in California but also in New Mexico."

Doing all this work isn't just about protecting your own home. "Firefighters have lost their lives trying to save homes, but if you have treated around your home and removed the fuels around your home it's much more defensible," Swetnam said. "Firefighters are much more likely to come in there to try to save your home, and it'll be a safer place for them to be while they're trying to put the fire out."

And that's not an abstract concern.

During the summer of 2017 I visited the Valle Grande in the Jemez Mountains. On the edge of the lush, green valley, there's a grove of towering pine trees.

The trees, many of them between 250 and 400 years old, comprise what's called the History Grove, and they offer a snapshot into what the forests of the Jemez Mountains looked like centuries ago.

Land management, as it's called, is made of up of meetings and programs, line-item budgets and public comment periods. And, sometimes, expensive lawsuits and bitter battles. But managing and protecting landscapes isn't an abstract exercise. It involves real people. People who wrangle with challenges and weigh in on decisions, and who sometimes make sacrifices.

In the History Grove the trees are spread apart from one another. The sun shines through the canopy, and green grasses run beneath the ponderosas. There are flickers and woodpeckers and other birds singing and squawking that bird watchers know but I can't identify. Driving here, we passed historic cabins built in the early to mid-twentieth century, including the one Sheriff Walt Longmire calls home on the Netflix series, *Longmire*.

In the winter I've seen elk lounging among the pines. On this trip Bob Parmenter, chief of science and resource stewardship at Valles Caldera National Preserve, pointed to a log where a bear had pawed through the soil underneath, looking for bugs.

Parmenter looked to the burnt ridge behind the grove and reminded me of the Thompson Ridge Fire. Ignited by a downed powerline at the end of May 2013, that fire burned nearly twenty-four thousand acres of the Valles Caldera and the Santa Fe National Forest.

Then Parmenter said something that would keep me up that night.

The Granite Mountain Hotshots saved this grove—and the cabins in it—before being assigned to the Yarnell Hill Fire in Arizona. The Hotshots were trained to fight both structural fires and wildland fires, Parmenter said, so they were the perfect crew to be at the Valles Caldera. Interagency Hotshot, or Type 1, crews are the most elite and experienced firefighting crews and are typically assigned to the most dangerous or unwieldy fires.

As the Thompson Ridge Fire approached the cabins in early June, the Granite Mountain Hotshots worked all night to protect them, Parmenter explained. They also lit backfires in the History Grove to save that iconic old-growth stand of ponderosas.

Shortly after leaving New Mexico, on June 30, 2013, nineteen of the twenty members of the Granite Mountain Hotshots died in the Yarnell Hill Fire. The crew was building a fire line to protect homes. But the winds shifted, and the fire blew up, trapping and killing Eric Marsh, Jesse Steed, Clayton Whitted, Robert Caldwell, Travis Carter, Christopher MacKenzie, Travis Turbyfill, Andrew Ashcraft, Joe Thurston, Wade Parker, Anthony Rose, Garret Zuppiger, Scott Norris, Dustin DeFord, William Warneke, Kevin Woyjeck, John Percin Jr., Grant McKee, and Sean Misner. Brendan McDonough survived the fire because he was stationed at a distance, working as a lookout.

Reporter Fernanda Santos wrote about the men and the Yarnell Hill Fire in her 2016 book, *The Fire Line*. "I was never truly aware of the job that wildland firefighters do, or the very important role they have in fighting fires but also treating the forests," Santos told me. Treating the forests can include reintroducing healthy fires—like the kind that burned through southwestern forests historically, nurturing stands like the History Grove—or allowing lightning-strike fires to burn naturally in certain places and under particular conditions. Those fires help clean up the forest to reduce the risk of the big, disastrous fires, like Las Conchas.

"I think a lot of people think that fighting a wildfire is like fighting an urban fire: you put the water on the flames and put them out," she said. "They don't understand that when they're watching on the evening news an airplane dropping slurry or water on the flames, down there most likely is a crew of twenty men and women working to stop those flames by carving trails in the forest, fire lines."

Many of those who live in the areas threatened by forest fires understand that human element.

Out on Highway 4 there was still a sign thanking the

firefighters who fought the Cajete Fire earlier in the summer of 2017. When the Cajete Fire started in the Jemez Mountains in June, Forest Service officials immediately called in a Type 1 fire-fighting crew because of the fire's complexity and its proximity to about three hundred homes and buildings.

Communities often react with gratitude during and after an emergency, posting signs along highways and inundating fire-fighter camps with supplies and donations. We are grateful when they save our homes and favorite places by putting their lives on the line for the things we value.

During prescribed or managed fires, we're not always that wel-coming, even though the work the crews are doing prevents those same destructive wildfires. We complain about the smoke or grum-ble when our favorite hiking trails are closed off during or after a fire.

Working on her book, Santos tried to understand not only what happened during the Yarnell Hill Fire, but what federal and state policies put firefighters in danger.

Writing about the Granite Mountain Hotshots has given her new perspectives. "Now, every time I feel very hot in the sum-mer"—she lives in Phoenix, where almost every day is hot in the summer—"I imagine how it is for people fighting fires right now," she said. "Or if I feel exhausted, I imagine what it must be like to work sixteen hours a day for fourteen days straight, carrying all that weight on their backs, and the tools in their hands."

Anyone willing to work in public service deserves respect, she said, including soldiers, police officers, and firefighters. And wild-land firefighters.

"They are a special bunch of people," she said, joking that most run away from reporters and don't want recognition or credit for what they do. "They are incredibly brave, they are incredibly strong, and they are incredibly important to our country," she said. That's especially true as fires get bigger in the western United States and expand beyond their historic range and outside the "normal" fire season.

In 2015 the nonprofit Forest Stewards Guild released a report on how communities in the Wildland-Urban Interface can best protect homes and neighborhoods from wildfires. The WUI—or "woo-eee"—refers to areas where homes and businesses meet or abut forests and undeveloped wildlands. Nationally, more than 10 percent of the US population lives within these zones—and developments are increasing all the time, despite the risks from wildfires.

The cost of fighting wildfires is rising, too. That's in part because more people live along the fringes of the forests—forcing firefighters to defend properties and communities when those bigger, more frequent fires occur due to warming. In the 1990s wildfire suppression comprised about 13 percent of the agency's budget. In 2014 it was more than half.

With so many homes within the WUI, conducting prescribed burns can be trickier than in more remote forests, like the Gila. Oftentimes, for example, people oppose projects, saying the smoke is a nuisance. Individual communities and families can buffer their homes from wildfire, said Alexander Evans, executive director of the Forest Stewards Guild. And agencies can undertake landscape-level projects, too—similar to the Southwest Jemez Mountains Collaborative Landscape Restoration.

"Not to be doom and gloom, but things will burn," he said, "and if we're not making the investment as taxpayers, we're going to pay the price as taxpayers in terms of rebuilding infrastructure." For Evans and others, water is an issue too. "If you destroy your watershed, you're going to be spending money to get water some other way," he said. "I'd rather pay, as a taxpayer, in advance and in a positive way that has these other eco-benefits, rather than trying to recover after the fire."

NO WATER HERE MEANS
NO HUNTING, NO RANCHING
STOP THE DRILLING
SanAugustinWaterCoalition.net

FIGURE 9. San Augustin Plains. Photograph by the author.

8. RUNNING DRY

GARRETT PETRIE AND Teri Farley moved to New Mexico about ten years before I visited them in the fall of 2017. They bought a house on five acres in the East Mountains—the back side of the Sandia Mountains, away from Albuquerque—because they liked being off the grid. Moving from Tucson, they felt like they understood water issues in the Southwest. "We asked a lot of questions," Petrie said. "We kept hearing things like, the wells really vary out here and you can get a good one, you can get a bad one."

They thought they had a good well. But immediately after the couple moved in, neighbors revealed problems with the well they shared. First they split the cost to replace a leaking storage tank. Some days they'd turn on the tap and water would trickle out. Other times the pipes would be dry, but then after a while the tank would refill. Then, after about two years, the well just quit producing. "We ran out of water," Petrie said. "It just stopped."

Initially they had water delivered—1,500 gallons cost eighty-five dollars and would last about two weeks. The cost added up, and they knew it was a just a short-term fix. Hooking into the local rural water utility, Entranosa Water & Wastewater Association, would solve the problem of their dry tap, but it also meant shelling out more cash—they had to build the infrastructure required to tie into the meter, which was more than four hundred feet from their home. They're still recovering from that financial commitment, Petrie said. "But you can't live without water."

The water company regularly sends reports showing it has a projected sustainable supply of more than a century, but Petrie wondered how realistic that is. "There's the pressure to grow and maintain a viable community. How is that feasible? How are we going to sustain that?"

This isn't just a problem for one home, or one well. Domestic wells are drying up across the Sandia Basin, a four-hundred-square-mile area across four counties, from Placitas to Tijeras and Sandia Crest to Edgewood. And a 2017 study by scientists at Stanford University and the University of California, Santa Barbara, shows this isn't unique to the East Mountains or even New Mexico: groundwater levels are dropping at an alarming rate across the western United States.

———

A STUDY CONDUCTED by Debra Perrone and Scott Jasechko looked at the data for more than two million groundwater wells across seventeen western states that were drilled from the 1950s until recently. As they wrote in their 2017 paper, published in the peer-reviewed journal *Environmental Research Letters*, of those, one in thirty wells is likely dry now.

In addition to learning that one in thirty wells no longer produces water, Perrone and Jasechko also found that agricultural wells are usually drilled to much deeper levels than domestic wells. "That implies that agricultural wells may be more resilient to drying out," Jasechko explained. "But on the other side of the coin, that implies that domestic wells may be more vulnerable to drying out than agricultural wells in these areas."

Jasechko also said they noticed two "hotspots" for drying in New Mexico: The Estancia Basin in Torrance County just south of Moriarty, "where a high proportion of wells have likely dried out," and near cities such as Portales and Clovis, which rely on the

Ogallala. In southeastern New Mexico, farms, cities, and outlying households rely almost entirely on groundwater. And the Ogallala, part of the High Plains Aquifer system, has been dropping for more than five decades due to overpumping.

Just a few months before their study was published, in the summer of 2017, the New Mexico Bureau of Geology and Mineral Resources released a report about the High Plains Aquifer that analyzed thousands of water-level measurements made over the past fifty years. They came to an alarming conclusion: "The High Plains Aquifer is rapidly being dewatered and its usable life is short. This is particularly so for large-scale irrigated agriculture using high-capacity wells."

Wells are drying up across the west, said Jasechko, making domestic water supplies vulnerable and possibly threatening farming in the western United States. Drying will affect food production and American jobs, he said. But it's also a social-justice issue: Not everyone can pay to dig a deeper well.

Perrone and Jasechko also found that well data varied from state to state. Some states track well data from the time a user applies for a permit to drill. Others, like Texas, rely on drillers to submit reports. In New Mexico the data is tracked from its point of diversion, but even that data isn't complete.

"In order to understand the vulnerability of drinking-water supplies and irrigation-water supplies, we really need excellent groundwater data," Jasechko said. "One of the results of our study is we learned there is more we can do in terms of collecting more water data. And the better data we have and the better information we have, the better able we are to manage these resources."

Just weeks after Perrone and Jascheko's paper was published, another was accepted for publication in the American Geophysical Union's peer-reviewed journal *Geophysical Research Letters*. In this paper, five scientists had looked at the impacts of climate change on groundwater recharge in the short term and in the long term. They noted that warming and climate variability affect groundwater systems both directly, through replenishment by recharge like rain and snowmelt, and indirectly, through changes in groundwater use through changes in demands.

Overall they found that northern portions of the western United States would receive more recharge, while the southwestern United States would receive less. In the near future, bracketed as the time between 2021 and 2050, replenishment in the Southwest will decrease by 4 percent, and between 2071 and 2100 it will decrease by about 9.5 percent. In their conclusion the authors noted that the Southwest is "already dry and stretched for water resource." Less recharge, they wrote, will "have significant challenges for managing water resources."

————

BEHIND A FIRE station on Frost Road in the East Mountains, county officials monitor four groundwater wells drilled to different depths. This area is "ground zero" for the problems in the East Mountains, said Philip Rust, a hydrogeologist with the Bernalillo County Water Resources Program. Readings from these four wells and others help him understand what's happening to the aquifer below.

But it's tricky. "Every individual well is so different that you can't speculate, unless you know that well or that area very, very well," Rust said. "The geology out here is so complex you can't generalize." Just as the mountains themselves crinkle, curve, and rise, so do the layers beneath the ground. "You don't know what you're going to encounter until you drill," he said.

Through Bernalillo County's water-level-monitoring project, Rust was also working with domestic well owners to monitor groundwater levels. "Since we've been collecting data, seven years now, aquifer levels have dropped about 1.8 feet per year on average in the East Mountain area," Rust said. That's an annual average for the entire area, which means some people—like Petrie and Farley—have seen drastic drops near them, while other wells might not have declined yet.

But in general, Rust said in 2017, aquifer levels are going down in the East Mountains. "A lot of these wells are ten, twenty, thirty, forty years old," he said. "So that's a lot of drawdown in each individual well." He explains that when people drill wells they usually stop once they hit water. If you drilled your well twenty or thirty years ago, it might not be deep enough today to reach the aquifer anymore. "Unfortunately, for a lot of people, their wells are critically in danger of not being able to produce."

About three hundred residents participate in the county's Water Resources Program, mostly in the East Mountains. And Rust is always looking for more people to join, whether their well is forty years old or brand new. To check a well's depth he uses a sonic device or a thick steel surveyor's tape. He usually does this twice a year, in the summer and late fall or early winter. Pinpointing those individual depths and changes helps him understand the basin better, and it helps people who rely on domestic wells know what's happening beneath their homes, too. "You can't make good decisions if you don't have good data," he said.

What scientists like Rust do know is that we're pumping water from the aquifer faster than rain and snow can replenish it. "If nothing else, it's a lesson about good resource management," Rust said. "Unfortunately, in the Sandia Basin, we have a limited resource and high demand. If we don't manage it effectively, we face the reality of critical loss, not having enough."

———

LIVING IN A COMMUNITY of about sixty homes just off the road leading to the Sandia Peak—about a dozen miles, as the crow flies, from where Garrett Petrie and Teri Farley live—Krista Bonfantine said she thinks about her water supply on a daily basis. She and her family were part of a cooperative that relies on water from a small spring on the east side of the Sandia Mountains. "I see mounting concern about water," Bonfantine said of the communities around her. "There are so many people affected, and so many people are seeing their friends and neighbors have to dig new wells, with huge price tags." When she was buying a house about fifteen years ago, they looked at some in the swanky community of Placitas at the north end of the Sandias. They were beautiful homes. But their wells were dry.

The small spring her community relies upon is Cienega Spring, which is part of what she called a "necklace of springs" at about the same elevation in the mountains. Initially people used it for irrigation. Then, infrastructure was built around it to capture water and pipe it downstream to the community.

On a sunny September afternoon, Bonfantine opened the heavy door protecting the spring and peered inside. Before my eyes adjusted to the dim light, it looked like water skimmers were skating and dancing on the surface of the water. Once I realized that I was seeing water percolating from the ground, my heart skipped.

But Bonfantine was worried about the spring. Like others in the Sandias, it's fed by rain and snowmelt. "The problem with a warming atmosphere is it's a thirsty atmosphere," she said. And there are increased human demands on the water, too. Currently she and others in the East Mountains are fighting plans by a development company to drill for more water in the area.

At that time Aquifer Science, a partnership between Campbell Ranch and Vidler Water Company, was seeking approval from the New Mexico Office of the State Engineer to drill for water in the Sandia Basin. "As the community faces increasing temperatures, water shortages, and we're already seeing diminishing supplies . . . seeing our state entertain a new appropriation of water . . . it just seems completely insane," she said.

"I think that every community in New Mexico is concerned about their water," she said, no matter where their water comes from or what their local issue might be. "As New Mexicans, some of our greatest resources are our natural resources, and water is necessary for all of our lives. We cannot in these times be in a position of having someone sell our water out from under us. We need to defend that."

These aren't new problems. And they shouldn't surprise anyone.

———

IN 2014 NEW Mexico state lawmakers set aside $100,000 to study the state's water supply. With that funding, five climate scientists, economists, engineers, and hydrologists from the University of New Mexico, New Mexico State University, and the New Mexico Institute of Mining and Technology pooled their expertise and other resources to meet some specific, statewide goals. They wanted to figure out how drought might affect rivers and reservoirs, groundwater supplies, and the state's economy—and then identify vulnerabilities and policy strategies.

After just a year of work, however, their funding was gone. Citing a drop in state revenue, the New Mexico State Legislature pulled funding for the group—known as the New Mexico Universities Working Group on Water Supply Vulnerabilities.

"There's no ax to grind with anyone, it's just a shortage of money and we're trying to prioritize money for public education and higher education," Sen. John Arthur Smith, D-Deming, chair of the Legislative Finance Committee and an advisory member of the Water and Natural Resources Committee, said in 2015. "That's the nature of the beast."

Between 2014 and 2015 the price of crude oil plummeted from almost ninety-nine dollars a barrel to fifty-two dollars. That drop, welcomed at the pump by motorists, was partly to blame for the legislature's decision to abandon the drought study. New Mexico, an oil and gas state, derives a portion of its revenue for government services from oil and gas drilling, making it sensitive to market fluctuations. That "sensitivity to oil has dramatically increased in recent years, magnified due to the record levels of production," state economists wrote in a memo. A $1 change in the price of oil produces a $10 million change in what flows into the state's major account—the general fund, according to that memo. With fewer dollars coming in, David Abbey, director of the Legislative Finance Committee, said policymakers had to make decisions. "The top education funding priority was instruction and general funding," Abbey wrote in an email. "Also there was an appearance of duplicate funding with many other water resource programs and initiatives."

Despite the lack of continued funding, the group's early findings are still valuable for water planners and citizens. The five-member study team began by studying southern New Mexico, where farmers and cities were worried about drought, dropping groundwater levels, and the US Supreme Court lawsuit Texas brought over the waters of the Rio Grande.

Droughts aren't a new phenomenon, of course. But by and large, in the twentieth and twenty-first centuries, communities survived droughts and water shortages by pumping groundwater. When the summer rains aren't enough to irrigate crops, farmers pump groundwater. The same thing happens when winter snows don't replenish reservoirs. Even when the state isn't experiencing drought, cities and farmers still pump groundwater. Because we can reach it—thanks to pumps, pipes, fossil fuels, and human ingenuity—we grab all the water we can. Because we can't see it, we think it's always going to be there.

But that's not the case.

In early 2015 Peggy Johnson, a hydrogeologist with the New Mexico Bureau of Geology and Mineral Resources at New Mexico Tech, told attendees at a meeting of the New Mexico Water Dialogue that pumping has depleted a resource that can take centuries to replenish. Between 1985 and 2010, water users depleted the groundwater beneath Doña Ana County in southern New Mexico by 2.5 million acre-feet—about the capacity of Elephant Butte Reservoir. Right now, she says, our rates of withdrawal—the amount of water we pump from beneath the ground—exceed the ability of those aquifers to recharge.

What does that mean? Basically, that New Mexicans in the future might not have groundwater to rely on as backup when rivers and reservoirs are low. Although in the past we've sustained ourselves through water shortages by relying on deep groundwater, she tells the group that "aquifers aren't going to be there to solve our water problems anymore."

Today, with the climate warming and scientists showing how higher temperatures are already affecting river flows in the arid West, groundwater could have been a cushion—a resource to help New Mexicans weather warming until we figured out more sustainable ways to support two million people on a handful of rivers and depleting snowpack. This reliance on groundwater, combined with our technological ability to grab every drop, perfectly illustrates the difficulty with which we're abandoning our cultural expectations.

During that water meeting in early 2015, people were still optimistic that the working group would continue and that its work would broaden out across the state. "The data isn't static," said Consuelo Bokum, longtime Water Dialogue board member, speaking to the assembled crowd. "If we're going to do planning, it's necessary to know this."

But the group didn't receive additional funding, and just a year later another troubling study came out, this one looking at eight aquifers in the western United States, including the Southern High Plains and Central High Plains aquifers, which supply water to New Mexico.

Published in the peer-reviewed *Journal of Hydrology*, the authors showed that withdrawals will continue exceeding the ability of aquifers to replenish with water from precipitation or irrigation and other water runoff. Mountain aquifers, which rely on snowpack to recharge, are also expected to decline. Typically, wrote the authors, studies have focused on global or local impacts rather than looking at changes on a regional scale. That poses problems for decision-makers, especially when aquifers cross boundaries and jurisdictions and are governed by agreements that assume the amount of available water will always remain constant. According to the study, "The key outcome is that existing information supports a 'wet gets wetter, dry gets drier' scenario. Southern portions of the western U.S. are likely to experience declines in recharge of varying magnitudes. Northern portions of the western U.S. may experience slight increases to modest declines."

Nationally, about 25 percent of water used comes from aquifers. In the western United States that number is higher, at about 40 percent. And in New Mexico? According to the US Environmental Protection Agency, 87 percent of New Mexico's public water supply comes from beneath the surface of the earth.

———

DRIVING ON HIGHWAY 60 across the Plains of San Agustin, it's easy to drift toward thoughts of the past. The floor of the valley cradled a lake during the Pleistocene, and historic windmills and stock tanks fleck the green expanse that stretches for some fifty miles, west of Magdalena and toward the Gila National Forest. But it's not the past that Catron County Commissioner Anita Hand worries about. It's the future.

More than a decade ago her brother and father spotted a legal notice in the newspaper announcing that the ranch next door planned to drill thirty-seven wells into the aquifer that provides water for the area. The Hands were not happy to learn about the new wells, which they feared would deplete their own. "My dad has spent most of his life building up what he has," Hand said of their family ranch. "He looks out there, sees all the work he's done in his eighty-four years might be for nothing because in the stroke of a pen it can all be taken away." After doing some calculations, the men were shocked to realize their neighbor's proposal called for pumping and moving seventeen billion gallons of groundwater out of the basin—annually.

Bruno Modena of Italy bought the eighteen-thousand-acre ranch in the 1960s. Since then, another rancher has leased the lands, running cattle for decades. But the owners now have bigger plans than cows: Augustin Plains Ranch, LLC, wanted to build a pipeline and pump fifty-four thousand acre-feet of water each year from the aquifer to the Albuquerque area. (When described geologically, the area is referred to as the Plains of San Agustin or the Valley of San Agustin. The ranch, however, is spelled differently: Augustin Plains Ranch.)

In the fall of 2017, when I spoke to Hand, the drilling application was in its third iteration and was once again pending before the New Mexico Office of the State Engineer. Under state law, New Mexico's waters, below- and aboveground, belong to the public and are held in trust by the government. The rights to those waters can be bought, like property. They can also be transferred or forfeited if they're not being used. To buy or transfer water rights, people or companies have to show the water will be put to "beneficial use."

According to its application, the company planned to sell water commercially in seven counties and to state and federal agencies. It also says it will provide water for the communities of Magdalena, Socorro, Belen, Los Lunas, Albuquerque, and Rio Rancho. The application didn't identify specific buyers who have committed to buy the water. But the application included a 2014 letter from Rio Rancho City Manager Keith Riesberg, who wrote that if the project is successful, the city would be "interested in discussing" moving water into the city's system. Nearby towns, like Magdalena and Socorro, actively fought the project.

Hand worried that if the state approves the project, it will open the door for private companies to make money off public water—water that belongs to the people of New Mexico.

But Augustin Plains Ranch Project Director Michel Jichlinski defended the project, painting it as a way for New Mexico to thrive through drought and dry times. "New Mexico and the American West need to develop more sustainable and environmentally sound water resources such as this project," Jichlinski wrote in an email. "We're confident it will be a trailblazer to a better future."

The project's public-relations contractor provided a fact sheet, too. In it the company said it planned to pump just one-thousandth of the estimated fifty-four million acre-feet of water stored within the aquifer. That estimate comes from a 1994 report. "Concerning the basin 'drying up,' it is a very dry area to begin with," according to the document. "The aquifer is several hundred feet deep. Plant and wildlife in the plains are supported by rainfall, not by the aquifer." It also said ranches and farms in the area will not be affected. "The plains themselves are sparsely populated and there aren't many operating wells." If other people's wells are affected and an "alternative solution" can be found, the company said it will pay for that. If there's no feasible solution, according to the company, "it is called impairment and the project cannot proceed."

The company's statement didn't satisfy local residents, or state Rep. Gail Armstrong. She said no one will know if their wells have been "impaired"—or dried up—until after pumping is already underway. "By that time, it's too late," she said. "Someone has got to stand up and make a decision on this," she said. That would be the state engineer. "It's just a big mess," Armstrong said of the proposal.

The state engineer, New Mexico's top water boss, should deny the application, Armstrong said, in part because the company still doesn't name an end user for the water it's planning to pump out of the aquifer. "In the meantime, all of my constituents are having sleepless nights," she said. "They're restless, wondering, 'Should we sell our place? Should we drill another well? What do we do?'"

Attorneys at the New Mexico Environmental Law Center, which represented local ranchers, residents, and the Gila Conservation Coalition in their protests against the application, were fighting this latest version of the application, saying the company hasn't made substantive changes that distinguish it from the one rejected by then state engineer Scott Verhines in 2012. At that time the agency called it "speculative" and said the company failed to say who would buy the water from the company. Augustin Plains Ranch appealed the denial, and the issue wound its way through the courts until the company filed a second application. Verhines rejected that one in November 2014.

Later that year company president Rich Radice filed a third application. In 2017 it was still pending before the Office of the

State Engineer—Tom Blaine, who Gov. Martinez named as a replacement when she ousted Verhines from the position.

Each time the company submitted its application, hundreds of people in Catron and Socorro Counties filed formal protests to the application. The original tally of protestants included about 1,000 people. In early 2017 the state's list included 604 "timely protestants." Law Center attorneys and state officials both say that drop in numbers may be due to confusion over the twenty-five dollar fee people must pay when protesting an application being considered by the state. And several tribal governments, federal agencies, acequia associations, and local governments in Socorro and Catron Counties also formally opposed the project.

Like Hand, Douglas Meiklejohn, the executive director of the Law Center, worried what approval of the application could mean not just for local residents but for other rural New Mexicans. "The main thing to understand is that if this can happen to an area like the San Agustin Basin, then it can happen elsewhere in the state," Meiklejohn said. "In terms of the possible precedent, it's a case that's important for all rural areas of the state."

Carol Pittman and her husband Ray have lived about a mile east of Datil for more than twenty years on land that borders the Augustin Plains Ranch. She and some others created the San Augustin Water Coalition, which they incorporated in 2008 to formally protest the application. "Ranchers don't know if they'll have water for their ranches, and residents don't know if they'll have water for drinking," she said. "Nothing can happen here while this is hanging over our heads. We're all kind of stuck." Pittman said the company's lack of communication with community members has been its "biggest stumbling block" over the past decade.

"They came up with this project and announced it, but they didn't talk to anybody about it, so when they did—and had that big meeting in Magdalena, where they were just telling them about the project and not asking—people really got razzed," she said. "If they had started out by talking with people, or asking people to be coming up with different ideas. Instead, it hardened everybody's resolve to get rid of them."

––––––

TOWARD THE END of September 2017, State Engineer Tom Blaine hosted a public meeting in the town of Socorro. Opponents of the water project filled an auditorium at the New Mexico Institute of Mining and Technology in Socorro, and New Mexico State Police officers kept an eye on things inside and outside the Macey Center.

During the meeting, state officials deflated emotions by keeping the meeting's agenda focused on process rather than on issues related to the application pending before the state and by limiting public participation to written questions and comments and not allowing attendees to speak before the crowd. Following a presentation on the application process, and after repeatedly reminding the crowd they could only answer questions related to the hearing process, two state officials spent about an hour reading the submitted questions aloud, sometimes having to decipher handwriting or recite what would have likely been impassioned comments.

Some questions were related to the process—if, for example, documents could be sent via email instead of the postal service. Officials couldn't answer a question about who would pay to drill new wells if existing domestic wells dried because of the development. Another unanswered question had to do with whether the aquifer had been mapped and modeled. A few people asked how the state will protect senior water rights or determine if those water rights had been harmed, or "impaired," by the drilling.

Some comment cards contained only statements, including one from a Catron County rancher who wrote that if the application were approved people "will be forced to defend ourselves." Randall Major, of New Mexico Cattlegrowers' Association, wrote, "This has been going on for ten years and should be put to an end now." Read in a controlled monotone by the state official, the fiery comment lost its impact.

Some comments, including those criticizing the "multinational corporation," called the meeting a "dog and pony show," and complaints about the lengthy hearing process elicited smatterings of applause from the attendees or the occasional, "Amen!" Some members of the mostly older crowd trickled out of the meeting, while others who stayed just mumbled their dissatisfaction.

At the end of the evening, Blaine returned to the podium. Two hours into the meeting only about sixty people were left in the auditorium—less than half the crowd from when the meeting began. "You can see that the focus was directed to the process, and the process that we have is governed by our state statutes," he said, adding that the state's water code is more than a century old. "It's been developed in the state to encourage applying water to beneficial use, but also to ensure that existing rights, senior rights, are protected, and we will, above all, make our decision in accordance with state statutes, because that is the fair thing to do for all of us."

On their way out Pete and Eva Dempsey of Datil called the meeting a waste of their time. They've already seen their water table drop by about ten feet in the past year, even without additional development in the area. If the application is approved, Pete Dempsey said, "people are not going to sit around. We don't want to see that kind of change." His wife cautioned him on some of his language. And because times are tense enough in America in the fall of 2017, I refrained from writing down all that he had to say.

Joan Brooks, also of Datil, echoed the Dempseys' opposition to the project. "My feeling about trying to develop this water: Why do we want to drive more people into a state where there is no water?" she asked. "We're a desert. It's just so insane."

Her words were nearly identical to Krista Bonfantine's description of the state's response to a company's plans to pump groundwater from the East Mountains. To the casual observer—like me—the two women appeared on opposite ends of the political spectrum. But their sentiment about water in New Mexico was exactly the same.

FIGURE 10. Rio Grande. Photograph by the author.

9. "HOT DROUGHT" AND DRY RIVERS

I COULD SMELL the mounds of dead fish before seeing them.

By then, the fish were desiccated. Most were lying in low spots along the riverbanks where they crammed together, taking refuge as the last of the puddles and rivulets dried. On a Sunday in early May 2018, the temperature was in the high 80s as a friend and I trudged up the sandy channel of the Rio Grande upstream of the town of San Antonio. I wondered what it must have smelled like two or three weeks ago, when the river first dried here.

The month before, when the Middle Rio Grande should have been rushing with snowmelt, New Mexico's largest river dried. The drying began in Bosque del Apache National Wildlife Refuge, spreading upstream more than twenty miles. Peering over the bridges in Albuquerque, you could see sandbars and slow water. Even in northern New Mexico, historically low flows were trickling through places like Velarde and Española, the result of not enough snow in the mountains this winter.

Traveling through an arid landscape susceptible to drought, the Rio Grande has often flowed in fits and starts. But until its waters were tamed in the twentieth century—by dams, canals, and increasingly sophisticated irrigation ditches—the river would also overflow its banks and swell across the wide floodplain. Those floods could wreak havoc on settlements and inundate farmland.

But they also nurtured native fish species, gave birth to the cotton-wood forests, and helped push the river toward the sea. Today the river is constricted and controlled, sucked dry by the demands of irrigators and cities and prevented from navigating new channels. As drought continues and climate change ramps up, the "Big River" is already a climate casualty in New Mexico.

The Rio Grande starts as a trickle in Colorado's San Juan Mountains and flows through the San Luis Valley and then on into New Mexico. Over the course of several million years, it carved out the Taos Gorge, and for centuries it has brought sustenance. Downstream of El Paso, Texas, the Rio forms the border between Texas and Mexico before it empties into the Gulf of Mexico. Along its journey the river provides drinking water to more than a million people and irrigates two million acres of farmlands, lawns, and orchards in Colorado, New Mexico, Texas, and Mexico.

Since the late 1990s the Middle Rio Grande has dried south of Albuquerque during the summer, when the river's flows are diverted into irrigation canals and ditches. In 2013 about thirty miles dried. Before the skies in September opened up and dumped rain for four days, people were bracing for water to disappear from the channel within Albuquerque. In 2003 the Middle Rio Grande ran dry for sixty miles. The next year almost seventy miles dried. In 2012 more than fifty miles.

Meanwhile, the river's channel through places like Hatch, Las Cruces, and Mesilla flows only when water managers release water stored in reservoirs and move it downstream to irrigators and cities like El Paso. Outside of irrigation season, which can vary in length from year to year, about one hundred miles of the Rio Grande in southern New Mexico is dry, save for when storms send flash floods down the channel. Indeed, as Frank Ward and James F. Booker wrote in their 2003 study of protections for endangered species, the Rio Grande below the reservoirs is "today largely a channelized water course serving primarily to convey water for offstream uses in New Mexico, Texas, and Mexico."

Of course, people can point to the fact that, historically, the river dried. In his 1954 Pulitzer Prize–winning book, *Great River*, Paul Horgan wrote, "The river commonly does not carry a great deal of water, and in some places, year after year, it barely flows, and in one or two it is sometimes dry." Arid-land rivers worldwide have flows that can wax and wane dramatically. That is, the riverbeds can be dry and sandy for a spell, then rage with floodwaters. But the morphology of the Rio Grande is different today. Until very recently in its geologic history, the riverbed wasn't broken up by dams and diversions. There were backwaters, oxbows, and dryland lakes that stored water and helped species survive. Thanks to humans tinkering with the system, those are absent now.

———

AS DRYING OCCURS more regularly, we need to understand what biological, chemical, and hydrological impacts are occurring, said Clifford Dahm, professor emeritus in the University of New Mexico's Department of Biology and an expert on intermittent and ephemeral rivers. "The aquatic creatures that live in the river, as it's drying and staying dry longer, are going to change," he explained to me. "There will be a shift toward completely different communities of fish, algae, invertebrates, and trees."

Now, we don't know exactly how quickly those shifts will occur or which species will survive, die, or recover. But when the water table drops to more than ten feet below the surface, we do know cottonwood trees struggle and then die, Dahm said.

We do know that in the Rio Grande Basin, warming will lead to a 4 to 14 percent reduction in flow by the 2030s and an 8 to 29 percent reduction by the 2080s. On the Colorado River—which New Mexico also relies upon—scientists have predicted a 20 to 30 percent decrease in flows by 2050. And a 35 to 55 percent decrease by the end of the century. Even on the Gila River in southwestern New Mexico, warming will decrease flows by about 5 to 10 percent due to decreasing snowmelt runoff.

These aren't isolated studies.

According to the 2017 Climate Change Special Report, global annual temperatures have increased by 1.8°F over the past 115 years, and the current period is "now the warmest in the history of modern civilization." The southwestern United States is among the most rapidly warming regions on the planet. Warming here is occurring at about double the global rate and will have profound impacts on the region's water resources, and by the end of the century, New Mexico could be 4 to 6°F warmer than it is today. Published in late 2018, the Fourth National Climate Assessment notes that human-caused climate change is already contributing to water scarcity in the southwestern United States.

At the same time that the earth continues warming, studies show that multiyear precipitation-driven droughts will continue to occur in the twenty-first century in the Southwest, and recovery from them will be "much more difficult" in the warmer climate. These findings are consistent with others on climate change and the US Southwest, which show the drying trend in the Southwest is predominately caused by a poleward extension of the subtropic dry zones; they also note evidence for this is difficult to separate from natural climate variability.

There is also a broad consensus among climate models that drying has already begun in the US Southwest and in northern Mexico. Natural climate variability "perturbs" a drier climate base state, which exacerbates drought risk, and in the US Southwest, models of weather-type frequency changes show up to a 25 percent decrease in precipitation. In addition, due to twentieth- and twenty-first-century reliance on groundwater, humans of the future will have less groundwater to rely upon. The loss of groundwater, plus the higher temperatures, will "likely exacerbate" the impacts of future droughts—leading to what the authors of a 2015 study say will be a "remarkably drier future that falls far outside the contemporary experience of natural and human systems in Western North America."

The warming trend has large implications for southwestern rivers. A 2015 study of 421 river basins in the Northern Hemisphere found that ninety-seven river basins—supplying water to almost two billion people—have at least a 67 percent chance of decreased snow supply. The authors of that study identified thirty-two basins most sensitive to changes in snowmelt, including the Colorado River and the Rio Grande. Under the two climate models that the authors used to simulate historical and future snowfall and rainfall, the risk of decline in spring and summer water availability in the Rio Grande Basin is between 95 and 100 percent. They wrote that continued warming-induced dryness, growing demands for water, and climate change–induced decreases in streamflows may force even large dry-land river systems—in particular, the Rio Grande—into "permanent hydrological drought."

Another study published in 2015, this one of twenty-four sub-basins of the Upper Rio Grande, found that the future annual volume of water is less than the pre–climate change model. Peak flow is already fourteen to twenty-four days earlier, and among the twenty-four sub-basins, daily hydrographs show higher streamflow in March and April but lower from mid-May through the end of irrigation season. As winters and springs continue to warm, the Upper Rio Grande in Colorado will see lower-than-historic flows in the late winter and early spring, according to a 2018 study. And new reconstructions of runoff ratio for the Upper Rio Grande Basin, stretching back to AD 1571, show that the declining trend in runoff that has been observed in the basin since the 1980s through the present is "unprecedented" in the last 445 years. Having strengthened over the past few decades, the authors of that 2017 paper write that temperature sensitivity implies that "future management vulnerability" will persist.

———

CLIMATE CHANGE IS already affecting the amount of streamflow in the Rio Grande that comes from snowmelt. "We see big changes in the winter and early spring," said Shaleene Chavarria, who studies the timing of snowmelt in the mountains that feed the Rio

Grande and its tributaries. "Big changes in winter temperature, increases in springtime temperatures, and decreases in stream-flow."

In 2018 the *Journal of the American Water Resources Association* published a study by Chavarria and David Gutzler that was based on Chavarria's graduate work in UNM's Earth and Planetary Sciences Department. Snowpack is the main driver of the Rio Grande's flows in the Upper Basin of New Mexico, she explained, and so she looked at annual and monthly changes in climate variables and streamflow volume in southern Colorado for the years 1958 through 2015.

She found that flows have diminished in March, April, and May. And not only is snowpack decreasing, snowpack melts earlier as temperatures continue to rise. This means flows have increased in the late winter and early spring and decreased later in the season, when farmers need it most for irrigating crops and orchards. The paper pointed out that as temperatures continue warming and streamflows drop during the growing season, more people will rely on groundwater pumping, further depleting already-stressed aquifers. "One thing that we really need to keep in mind is that conditions in the basin are changing—we know from the paleoclimatic records and stories from the past—and we have projections of what could happen in the future," she said. "We need to take those together, and we need to prepare, because ultimately what we do today and in the coming years is going to affect future generations—mainly our children and our grandchildren, so we need to take care of what we have now."

Like many New Mexicans, Chavarria has seen the impacts of climate change firsthand on her community. In northern New Mexico, the Pueblo of Santa Clara's forests and watershed have been hammered by fires, floods, and forest die-offs. "I come from Santa Clara Pueblo, and our traditions revolve around the seasons and revolve around the water that we get from the river," she said. "We've seen a lot of changes in our area especially with the fires that have impacted our main watershed."

In 2011 Las Conchas devastated Santa Clara Creek and its watershed. "We had that major fire and then right after the fire, we had debris flows and flooding, which pretty much added to the effects that the fire had on our community, and so it's still recovering."

Being a member of Santa Clara informs her work as a scientist: "I don't want to speak for everybody, but my perspective growing up is that everything works together, goes together, and your decisions affect everything else," she said. "So, for example, if you degrade the environment, it's going to have an impact on future generations . . . it's the same concept in the water cycle, everything working together."

Her coauthor and graduate adviser, Gutzler, said the study has one major takeaway for policymakers: snowpack is becoming a less reliable source of streamflow in New Mexico's rivers. Even in snowy years the region's warmer springs melt the snowpack faster and earlier than in the past. "New Mexico is a variable climate, so for millennia we have seen wet periods and dry periods on the order of decades. What's not so normal, by historical standards, is how warm it's been," he said.

"Last year, in 2017, we had above-average snowpack up almost to the point of peak snowpack, and it just melted away," he said. "So even in a year where snowpack looked terrific, we ended up with nearly average streamflow by historical standards." And that's a warning of things to come. "Thinking long-term, we still don't have a really good idea of how much streamflow to expect as a new normal—the interplay between precipitation and temperature is different in a warmer climate, and we're still sorting out what's likely to happen in the rivers as the climate gets significantly warmer. I think the paper we just wrote is a contribution to that, but there's an awful lot of work to be done to narrow down the uncertainties of what sort of future climate we ought to expect and plan for."

The Southwest isn't going to run out of water, Gutzler said. But we do need to make changes. "The ways we've managed water in the past may not work to provide all the water that people think they need in the future," he said. "And the sooner we plan for that, the more likely we are to end up managing water in a way that does satisfy people's needs."

———

AT A CONFERENCE in Santa Fe at the end of April 2018, climate scientist Jonathan Overpeck spoke about "hot drought"—dry periods that are warmer than past conditions—and megadrought. Looking back at the tree-ring records, the longest drought—or megadrought—on the Colorado River and the Rio Grande headwaters was fifty years long, around AD 200. Warming increases the chances of a megadrought, Overpeck said: of a twenty-five-year-long megadrought by 17 percent, of a fifty-year-long megadrought by 8 percent. "If you're really planning for the future," he said, "this is what you have to plan for." Future warming is a "sure bet," he said. "There isn't a bet you could make that's more guaranteed to come true."

Depending on which emissions scenario modelers use, warming in the southwestern United States will mean a temperature rise of 4 to 7°C by midcentury. Surface-water risks are exacerbated, he added, by unsustainable groundwater use. "If we continue business as usual"—referring to greenhouse-gas emissions—"flows will be knocked by 50 percent by the end of the century," he said of the Colorado River. "That's the cost of global warming for the Southwest."

Overpeck thinks that the Southwest, including New Mexico, can become known for developing water strategies to deal with scarcity. "That's the future economy of the Southwest, not sticking our head in the drier and drier sand as the news gets worse," he said. "There will be wet years—and when we get a good year, don't just allocate that water to a new user, put it underground." States like Arizona, Nevada, and California already store water underground, instead of in reservoirs, which lose more and more water to evaporation as the region warms. "The other really good news," he said, "is we know what the problem is."

Almost a week after hearing Overpeck's talk, I returned to the dry riverbed of the Rio Grande, bringing along a friend who moved here from Germany. Walking through the cottonwoods and Russian olives, when we spotted the dry channel, he asked, "That is the Rio Grande?"

That, I said, is the Rio Grande. And we hopped down into the channel, sand filling our shoes.

FIGURE 11. Elephant Butte Reservoir. Photograph by the author.

10. "IT'S NOT DOOMSVILLE YET"

ON THE DOWNSTREAM side of Elephant Butte Dam, just north of the town of Truth or Consequences, three US Bureau of Reclamation employees navigated a stairwell above the Rio Grande, passing scat from the ring-tailed cats that like to hang out here, and entered through a door into the three-hundred-foot-tall concrete dam.

Built in the early twentieth century, Elephant Butte Dam holds back water stored for farmers in southern New Mexico, Texas, and Mexico. At full capacity the reservoir is about forty miles long and can retain more than two million acre-feet of water.

Jesse Higgins, an electrician who manages the power plant at the dam, went in first and flipped on the lights, which flickered and fired up after a few moments. Labyrinthine tunnels burrow throughout, and water drains along the sides of the narrow, elevated path. Inside, it's easy to imagine what the world was like in 1916, when the dam was completed. The Civil War had been over for half a century—nearly comparable to the time between the Vietnam War and now—and the Mexican Revolution was ongoing. Since 1916 there have been world wars and shifting alliances, medical and technological breakthroughs. Humans have visited the moon and landed a rover on Mars. Our understanding of the earth and humanity's impacts upon it have changed as well.

During that time comparatively little has changed when it

comes to how water is managed in New Mexico. The Rio Grande Compact, which divides water among Colorado, New Mexico, and Texas was signed in 1938. And New Mexico's water laws today are still based on codes that the territorial legislature passed in 1907. But the same tactics and strategies that may have helped New Mexicans weather dry times over the past century won't keep working. And perhaps no place in the state offers such a stark reminder of that fact than the reservoir behind this dam. After a dry winter and hardly any snowmelt this spring, in mid-September 2018, Elephant Butte Reservoir was at 3 percent capacity, storing 58,906 acre-feet of water. "There was no spring runoff this year. We started this year at basically the point we left off at last year," said Mary Carlson, a spokesperson for the US Bureau of Reclamation, which operates Elephant Butte Dam. The federal agency runs the Rio Grande Project, which stores water that legally must be delivered downstream to the Elephant Butte Irrigation District (EBID), Texas, and Mexico.

The winter's dismal snowpack broke records in the headwaters of the Rio Grande. Without spring runoff, by February reservoir levels around the state—including at Elephant Butte—were as high as they were going to be this year. "We had some help from the monsoons," Carlson said, "but not as much as we wanted, where we wanted."

Many spots around New Mexico reveal signs of drought and climate change: the puny flows of the Rio Grande, the fire-ravaged forests of the Jemez Mountains, the crispy rangelands of the northeast. But Elephant Butte Reservoir in 2018 offered perhaps the starkest reminder that keeping up with the changing climate may require questioning long-held ideas of how water is managed and shared, how we think about rivers and reservoirs, and even who we consider our friends or foes.

———

FOR FARMERS IN southern New Mexico, the year 2018 "really

stung," said Gary Esslinger, manager of EBID. Less than forty-five thousand acre-feet of water flowed via the Rio Grande into Elephant Butte, he said—the lowest recorded inflow since the dam was built in the early twentieth century. "Whatever there was didn't get to Elephant Butte," he said. "The Middle Rio Grande, that river was drying up way too early." Beginning in early April, it began to dry south of Socorro and upstream of the reservoir.

If you've eaten chile from Hatch or pecans from Mesilla, fed alfalfa to your horses or poured milk from a New Mexico dairy into your coffee, you've consumed water that EBID's farmers divert from the Rio Grande and Elephant Butte or pump from the aquifer. The district has over ninety thousand acres of irrigable land and about eight thousand members, about half of whom own two acres of land or less, Esslinger said. This year, about seventy-five of the district's acres were in production, and farmers received a ten-inch allotment of water from the Rio Grande Project. A normal allotment is thirty-six inches, and in 2017—after a robust snowpack in the mountains—they received twenty-four inches. As in the past, farmers supplemented their irrigation supplies from the river by pumping groundwater. That's something farmers have done for decades, increasingly so since 2003. Farmers with larger landholdings will fallow some fields, not planting some fields so they can move what surface water they have closer to their irrigation wells, Esslinger explained. Then they can "stack" both surface water and groundwater on those lands.

But Esslinger let out a long, deep sigh when asked what will happen next year. Farmers can hope that the forecasts are right, that conditions in the Southwest will flip from La Niña to El Niño, bringing moisture to the region, he said. "But I'm dealing with 'La Nada,'" Esslinger says. "I have to face reality." Watching the reservoir empty out this year makes farmers feel like they are running out of water, he said. At the same time, they're uncertain about how long their groundwater supplies will last, even though the district tries to monitor groundwater levels and has hired a full-time groundwater specialist.

"We're not cratering; it's not Doomsville yet," he said in the fall of 2018. "But we've got to find another source." People can pray for rain and snow, he said, but the challenge is finding a long-term, consistent water source. And western states, including New Mexico, don't have that.

With improvement unlikely, Esslinger said he's started considering more radical solutions—like whether western states could share the cost of a canal that would move water from the East, from someplace like the Mississippi River. "People think I might be crazy, but I think we should start looking at it," he said. "I don't think we can continue to keep playing this game of predicting and forecasting: we need to find some water and get it over here to the West."

Farmers face other challenges, too, including the growing expense of pumping groundwater and an "insurmountable" number of regulations, he said. It's also hard to find workers to hand-pick crops like chile and onions thanks to changes in immigration policy. Then there are the market pressures. In 2018, he explained, farms around Yuma, Arizona, and southern California flooded the market with onions, forcing New Mexico farmers to sell theirs at a lower price. Cheaper alfalfa comes up from Mexico, he said. And even chile farmers have taken a hit. "In Mexico, they can grow chile and jalapeños much cheaper than we can grow it here, because of the labor, and they ship it here," he said. "Then, when our chile is ready, the market we could have had has already been flooded by a lower-cost chile."

Meanwhile, as the average age of farmers in the West keeps rising—most are in their fifties, sixties, or seventies—Esslinger questions who will farm New Mexico in the coming years. "It would be like taking your life savings to Vegas and gambling: what young farmer would want to do that?" he asked. "Or, if you're a farmer from Iowa or someplace else, where you grew up with a plentiful amount of water and rainfall to grow your crops, why would you come here?"

Farmers in southern New Mexico have yet another problem:

uncertainty over a lawsuit moving through the US Supreme Court. In 2013 Texas sued New Mexico and Colorado, alleging that New Mexico failed for decades to send its legal share of Rio Grande water downstream by allowing farmers in southern New Mexico to pump groundwater from near the river. Texas filed the lawsuit after New Mexico sued over a 2008 operating agreement between the US Bureau of Reclamation, EBID, and Texas water users. In early 2018 the Supreme Court allowed the US government to intervene in the case against New Mexico. That means New Mexico is squaring off against Texas *and* the federal government.

Although EBID lies within the boundaries of New Mexico, for the purposes of water and compliance with the Rio Grande Compact of 1938, it's more closely aligned with Texas. That's because under the compact, New Mexico doesn't deliver Texas's water at the state line. Rather, water goes to Elephant Butte Reservoir, about a hundred miles north of Texas. From there, Reclamation delivers it to farmers in both southern New Mexico and Texas. "We get a lot of harassment and bad press, with people saying, 'Why can't you just agree with [the state of New Mexico]?'" Esslinger said. "But when it comes to water accounting and the federal accounting of water through [Colorado, New Mexico, and Texas], we're in Texas."

All that confusion and uncertainty just makes things harder for farmers, according to Esslinger. "There are so many lines in the sand that have been crossed by our own officials in the state that it makes it very difficult to sit in a room and even try to talk about settlement or negotiation," he said. "Everyone is fearful of what they might lose, so they have fortified their positions." People on opposite sides of the suit can't even visit with one another, he said. And they certainly can't plan for next year, never mind the longer-term future.

Meanwhile, as drought lengthens, water managers are refining their models and developing new technologies to manage water and also do things like reduce evaporation from reservoirs, said Reclamation's Carlson. Agencies, irrigation districts, hydrologists, and stakeholders are in constant communication with one another, moving water and trying to work together in new ways.

Along with partners, the federal agency also tried to keep as much water as possible flowing in the Middle Rio Grande in 2018, said Carlson. At the end of summer, for example, Reclamation leased twenty thousand acre-feet of water from the Albuquerque Bernalillo County Water Utility Authority to keep the river running through the city through the end of the year.

Meanwhile, the agency will continue refining its tools and technologies for modeling, forecasting, and water delivery to figure out how to make it through next year and the years after that. "As you get stressed, you have to look for those outside-the-box ideas," said Yvette Roybal McKenna from Reclamation's Water Management Division. "We have to find the optimum path so we can move forward and adapt." She said she can't accept a future where the project fails to deliver water. "We're going to do everything we can."

Reclamation has also been studying climate change and its effects on the Rio Grande Basin, which supplies drinking and irrigation water for more than six million people. Between 1971 and 2001, average temperatures in the Upper Rio Grande Basin increased by an unprecedented 0.7°F per decade, or double the global average. And they're expected to rise within the basin by an additional 4 to 6°F by the end of the twenty-first century. Those rising temperatures will cut the amount of water flowing into the system as well as the timing of those flows, according to a 2013 report from Reclamation about the impacts of climate change on the Upper Rio Grande Basin. At the same time, more water will evaporate from reservoirs. And plants—forests and crops—will demand more water to survive. All of these factors together, according to the report, "are expected to cause significant changes in the available water supply and demand."

A 2016 Reclamation report also notes that the river's flows are already insufficient to meet the basin's water demands, and the basin already experiences water-supply shortages, even without the effects of climate change.

And while it's alarming to see reservoirs like Elephant Butte drop so low, they have done their job: Storing water from past years allows people to survive dry years like 2018, Carlson said. Despite the year's historically dry conditions, Reclamation delivered about 60 percent of a full supply of Rio Grande Project water in 2018. "We live in the desert and are more and more dependent on reservoirs," she said. "This is the year that reservoirs were built for; our reservoirs are doing what they were meant to do, and this year, Elephant Butte performed like a champ."

As for next year? "We're all on the edges of our seat," she said. "Waiting to see what's to come."

––––––

ON THE TAIL end of a good snowpack in southern Colorado and northern Mexico toward the end of 2017, UNM's David Gutzler told a crowd of people at a November climate-change conference in Albuquerque that he was "openly skeptical we'll ever be able to fill Elephant Butte Reservoir again."

Since Reclamation completed the reservoir in 1916, its levels have fluctuated—from highs in the 1940s to lows in the 1950s, '60s, and '70s. Many New Mexicans are familiar with the wet period that lasted from 1984 through 1993; between 1980 and 2006, the state's population increased by 50 percent. But then the region was hit with drier conditions—and increasing temperatures. Current droughts are not just caused by a lack of precipitation but a rise in temperature. "It is a lot warmer here now than it was a generation ago," said Gutzler—about 3°F warmer. And by the end of the century, New Mexico could be 4 to 6°F warmer.

In 2010 he and a colleague published a paper about rising temperature trends in central New Mexico. "By the end of the twenty-first century, Albuquerque's temperature will be the current temperature of El Paso," he said, noting that the vegetation in the Franklin Mountains flanking the west Texas city look very different from the Sandias, which still have aspens and conifers in higher elevations. "The climate in El Paso is significantly different from Albuquerque even though we get roughly the same precipitation."

Pointing to the screen behind him, Gutzler drew attention to the red temperature curve, running from 1935 to 2015, with projections beyond. "That red curve is headed up," he said. "And the choices we make will determine how much higher that will go." Given continued emissions of carbon dioxide and other greenhouse gases, warming will continue to accelerate globally, and in New Mexico.

The biggest foreseen impact on New Mexico will be on snowpack, Gutzler said. "We will still have wet decades," he said. But they won't boost reservoir levels or recharge groundwater as efficiently. Soils will be dry. Water demands will be high, from cities, farmers, and vegetation. And reservoirs and ecosystems alike won't be able to catch up on water. "We don't have a choice whether to adapt to warming," Gutzler said. It's happening—and it's going to continue happening. New Mexicans can choose, however, to reduce emissions. "If we want to, we can choose a lower emission rate and a 'softer path' to warming," he said at the end of 2017. We can think globally and long-term, he said, and join other countries to reduce greenhouse-gas emissions.

———

"HISTORICALLY, PEOPLE TEND to listen to what they want to hear, rather than what they need to hear: What they need to hear is that our laws do not reflect hydrology, and our hydrology is changing for the worse, and if we do not manage it, it will manage itself," said Phil King, a consultant to EBID and an expert on hydrology and the relationship between surface water and groundwater in southern New Mexico. "I would much rather correct the system ourselves through management than let nature do its cold, hard reality fix," added King.

For roughly a century the district's farmers have supplemented irrigation water with groundwater. Without it, they would not have survived the drought of the 1950s. But they pumped during the wet years, too, including throughout the 1980s and '90s. Then, beginning around 2003, about four years into the Southwest's current drought period, pumping ramped up even more. That's a problem, especially in the Rio Grande Valley, where river water recharges the groundwater and pumping water from the aquifer makes it even thirstier for river water.

With both the surface water and the groundwater strained, the system suffers a double whammy, King said. That causes a positive feedback, or what King calls a "death spiral." Even though scientists, engineers, hydrologists, and farmers know the two are intertwined within the same system, in New Mexico groundwater and surface water are managed separately. King calls that "hydrological folly." "We've got some major rethinking to do with New Mexico water law: Status quo is not an option," he said. "I think what people need to understand is we are facing conditions that mankind has not faced here before."

And the only way to reverse that death spiral is to use less water.

King said one way is to formalize a fallowing system that allows cities, factories, and businesses—in Las Cruces or in burgeoning border cities like Santa Teresa—to pump groundwater if they pay southern farmers who own surface-water rights to fallow their fields. Another way is for farmers to reduce their irrigated acreage and grow higher-value crops.

It's clear that any real solutions to cut water use must focus on agriculture. That's because farms use 75 percent of the water in the Rio Grande Basin. Cities can implement conservation measures, and people can reduce their household water use, King said, but the overall savings are minimal. Even finding "new" sources of water to add to the system—like capturing stormwater runoff or desalinating brackish water—will only add tens of thousands of acre-feet, King said. That doesn't come close to making up for the amount of water drought and climate change depleted from the system.

In King's ideal world, water-management schemes would reflect the connection between surface water and groundwater. And water management wouldn't get blown off course by political winds. "I think the handling of water policy, in terms of both promulgation and implementation, needs to be de-politicized," he said. "It needs to be based much more on science, hydrology, and the hydraulics of the system, rather than on politics." Then, rather than each sector—agricultural, municipal, and industrial—fighting over every last drop of water, solutions could emerge. And so, too, could changes that protect the river and groundwater system, the economy and people's futures.

———

ONE IDEA TO keep more water in canals and pipes, as well as in the Rio Grande itself, is to stop storing water at Elephant Butte. "Keeping water in Elephant Butte is a practice I think is out of date, and not wise," said Jen Pelz, an attorney for WildEarth Guardians. Located in southern New Mexico—an arid environment that keeps getting warmer—Elephant Butte Reservoir loses an enormous percentage of water each year to evaporation.

Rates of evaporation vary depending on humidity, wind, radiation, temperature, and the amount of water actually in the lake. According to a 2004 report from New Mexico State University, evaporation from Elephant Butte can be up to one-third of the average inflow each year. Between 1940 and 1999, when inflows to the lake ranged from 114,100 acre-feet to more than 2.8 million acre-feet per year, annual evaporation averaged about 250,000 acre-feet of water. Warming will only accelerate Elephant Butte's evaporation rate—by another 10 percent, according to Reclamation's 2016 report.

That means it's time to change where water is stored on the Rio Grande, said Pelz.

WildEarth Guardians wants the National Academies of Sciences to evaluate existing reservoirs in the basin and run models of how the system would function if water were stored in different places, such as in upstream reservoirs with lower evaporative losses. Storing Rio Grande Project water—the water in Elephant Butte that Reclamation has to deliver to EBID, Texas, and Mexico—in higher-altitude reservoirs would could save between forty thousand and eighty-five thousand acre-feet a year from evaporating, according to a report from WildEarth Guardians called "Rethinking the Rio."

Changing where water is stored would mean renegotiating parts of the Rio Grande Compact of 1938. And since federal laws passed during the twentieth century lay out the rules for reservoir operations and water storage, Congress would need to take action. "People have been talking about reservoir reoperation for a long time, but no one talks about how you do it," she said. "You have to deal with the compact, deal with the reservoir reauthorizations, deal with accountability along the river." If water were stored higher in the system, for example, downstream users would need to know their upstream neighbors weren't diverting their water unfairly.

Making these monumental changes demands building trust and relationships within the watershed, said Pelz. But New Mexico's vulnerability to climate change—revealed so clearly in 2018—should motivate everyone to start doing things differently. "For the middle valley and in the south, [managers] delivered all the water for irrigation this year," Pelz said. "And if the reservoirs can't be filled up over the winter, there will be no water for next year."

That's a crisis, Pelz said, for the Rio Grande and for the people who depend upon it. "Taking concrete steps to do something different means sacrifice: The reality in New Mexico is there are going to be sacrifices, areas that get dried up, and people have to change the way they make a living," she said. "That's the reality of the climate-changed world we live in."

———

MEANWHILE, THE US Supreme Court battle between Texas, New Mexico, and the US government over the waters of the Rio Grande marches onward. At a meeting at the end of August 2018, the special master assigned to the case by the Supreme Court set some new deadlines: the discovery period will close in the summer of 2020, and the case is slated to go to trial no later than that fall.

If New Mexico loses the case, Texas could seek damages of up to $1 billion, compensation for the more than three million acre-feet of water the state says it should have received over the course of more than a half-century. Not only that, but New Mexico could be forced to curtail groundwater pumping throughout the Rincon and Mesilla valleys, the hundred-mile stretch from below Elephant Butte to the United States–Mexico border.

Looking at Elephant Butte Reservoir in the fall of 2018, it's easy to think that New Mexico doesn't have any water to send downstream. But it's the water below ground that has been a point of contention between the two states. Texas has long complained that by allowing farmers to drill wells alongside the Rio Grande, New Mexico has siphoned off water that is hydrologically connected to the river and should be flowing downstream to Texas under the Rio Grande Compact of 1938. When the case ended up in the Supreme Court, EBID tried to intervene and become a party to the case. As Esslinger explained, the irrigation district is in a sort of limbo: it's in geographic New Mexico but "compact Texas."

The Supreme Court denied that intervention, but the irrigation district did file an amicus, or friend of the court, brief. And in 2018 the new special master on the case, Judge Michael Melloy, granted both EBID and El Paso No. 1 an "enhanced level of participation," said EBID's attorney, Samantha Barncastle. That status allows the districts to take part in the ongoing discovery and deposition processes.

Still frustrated that it's on the fringes of the legal dispute, even though it could affect their farmers so intimately, the district is increasingly focused on how agriculture can survive into the future—given constraints on surface waters—due to drought and warming. "We're not going to have the surface water we had in the '80s and '90s, so as an irrigation district we need to find out a way to survive to benefit the economy in southern New Mexico," she said. "It's nice to rely on the surface water, but it's not materializing every year like it used to, or it's different than it used to be," she says. Sometimes precipitation comes later in the year, as rain instead of snow, or it falls below the reservoir and can't be captured for storage.

"At this point, everybody needs to understand it's absolutely necessary to rely on the groundwater, but we have to do that in a responsible way," Barncastle said in 2018. "What EBID would like to see happen is [to have] responsible limits on how groundwater can be used." EBID's farmers are looking for solutions, she said, not litigation. And all the water users need to figure out what it means to have a resilient aquifer in southern New Mexico. Municipal and industrial water users rely on groundwater, too, she pointed out. And if groundwater is the "savings account"—the water people draw upon when surface water isn't available—everyone needs to know their spending limits.

"We're looking at, how do we make sure this area survives into the future, assuming we aren't always going to have the surface water available," said Barncastle. "Something's got to give, so what are the give points, and who is going to participate? Our position is it's not just the farmers that should be cutting back and getting responsible about water use—it's everybody."

———

IT'S SAFE TO say that 2018 was a tough year in New Mexico—for farmers, but also for anyone paying attention to rivers and reservoirs, and anyone paying attention to the world around us.

That fall I gave a handful of presentations about water, climate change, and politics in quick succession over a few weeks. Flashing the requisite slides on the screen—a dry Rio Grande in April, woefully low reservoirs in summer and fall, graphs showing temperature increases over decades—I watched peoples' faces in the audience.

I watched their shock over emptying reservoirs. Their grief at the photo of hollowed-out fish at the edge of the sandy channel, where they took refuge until the last puddles dried. I watched the older white men whose faces were the physical manifestation of the messages I sometimes receive: "What does she know?" And I thought for a long time afterward about the silver-haired woman who teared up when she mentioned her brand-new grandbabies. She turned from me as she wondered aloud about their future. Then she stopped talking. And I didn't know what else to say.

In October 2018 a special report released by the Intergovernmental Panel on Climate Change explained we have about a decade to prevent catastrophic and irreversible warming—but only if we choose now to change our energy systems. For more than a century we've extracted fossil fuels from the ground and burned them to power our economy, drive our cars, heat our homes, and fuel our war campaigns across the world.

It turns out that burning up millions of years' worth of carbon in only a century or so has consequences.

Like many of the people whose faces I watched, I felt the tug of despair. And over the past few years I'd even tried to stay my heart against landscapes that previously offered solace.

It's not emotionally restorative, after all, to witness swaths of conifer forest degrade from green to brown in a matter of seasons. To watch the Rio Grande turn from mud to sand in April, when snowmelt should have been swelling over its banks. To see a favorite little canyon—where every year I watch tinajas fill with water, then tadpoles and snakes—stay mostly dry, even after monsoon season.

That's not to mention wildfires, expanding dunes, oil fields and natural-gas flares, changing ecosystems, shifting migration patterns, and disappearing wildlife species.

It is my job as a journalist to bear witness. And to report these changes.

But over the past few years, I started failing to notice beauty beyond the scars on the landscape. And I often found it harder to connect with humans, even those I love the most.

Then I busted my rotator cuff. It was impossible to drive without contorting myself to steer and shift with one hand. So I started walking around my neighborhood instead of driving to hiking trails, where it's my habit to actively avoid anyone else who might be out there too.

In late October 2018 I started walking an hour-long loop that left me standing at the corner of a field full of sandhill cranes just before sunset. I would watch the Sandia Mountains turn pink and then purple while the cranes lifted off in small groups, heading west toward the river for the night. Sandhill cranes rely upon the Rio Grande for shelter at night and to navigate, as they have for thousands of generations, north in the summer, south in the winter.

As weeks passed, more people clustered along the field each day. Almost always we would each stand quietly, waiting for the birds to run a few skipping steps and take off toward the west, to settle in groups in the river for the night. We would smile up at the sky as their wings powered them toward the river a half-mile away.

Transformed, we would each head back to our families and loves, responsibilities and challenges.

It wasn't until one November weekend that the pre-twilight ritual reminded me to excavate a picture my daughter had drawn in elementary school. In yellow highlighter and ballpoint pen she captured a flying crane—its neck outstretched and feet dangling—after a visit to Bosque del Apache National Wildlife Refuge.

I kept it, even though the ink has faded and I can barely make out the words she wrote in the top corner: "cranes are our fall gods."

Year after year, when the birds migrate here to winter along the Rio Grande, their calls remind me—and likely anyone who hears them—of the past. Their whirring croaks have a primordial quality, and even in the middle of Albuquerque, it's possible to close your eyes, listen to their calls, and imagine the past.

And, just maybe, imagine the future too.

One evening, a woman who looked to be in her seventies spoke as she walked slowly up the dirt path. Her face resplendent, she smiled at me: "We are the luckiest people on the planet." All I could do was grin like a mad fool and agree with her.

FIGURE 12. Tres Piedras. Photograph by the author.

AFTERWORD

THE IMPACTS OF climate change are more and more obvious all the time: while scientists focused on sea ice melting in the Arctic, anomalies in sea-ice extent and behavior started popping up in the Antarctic. As I wrote this in early July 2019, a heat wave in Alaska was setting records, and almost five hundred firefighters were battling the blazes that started igniting in June. Worldwide we're seeing more and more problems we can't manage. In addition, the science of climate change continues at a pace that makes it hard to keep up as a beat reporter, never mind within the context of a book, which takes years to write, assemble, edit, and produce.

Regularly, new studies point toward what's happening across the planet, and what's forecast to come. Some even upend what we think we know about the past. In 2019, for example, a study in the peer-reviewed journal *Nature* revealed signs of human-caused climate change in the past, too. Drawing upon computer models and long-term global observations, the study showed the "fingerprint" of drought due to warming from greenhouse-gas emissions in the early twentieth century. The researchers identified three distinct periods within their climate models: 1981 to present, 1950 to 1975, and 1900 to 1949.

In that initial time period, during the first half of the twentieth century, "a signal of greenhouse gas–forced change is robustly detectable," they wrote. Both various observational data sets and tree ring–based constructions of drought over the past millennia "confirm that human activities were likely affecting worldwide drought risk as early as the beginning of the last century."

Drought is complicated, explained Benjamin Cook, with the NASA Goddard Institute for Space Studies and Columbia University's Lamont-Doherty Earth Observatory, and one of the study's coauthors. It can be measured in different ways, including by precipitation, streamflow levels, agricultural drought, or runoff levels. Their study examined drought in terms of soil moisture, and they relied upon tree ring–based reconstructions of the Palmer Drought Severity Index (PDSI). That index uses temperature and precipitation data together to estimate dryness. And "drought atlases"— collections of tree-ring records that reconstruct the PDSI for North America, Europe and the Mediterranean, Mexico, Monsoon Asia and Australia, and New Zealand—allowed them to look at records stretching back six hundred to nine hundred years, depending upon the region.

Detecting the signal of climate change on the global hydroclimate in the first half of the twentieth century surprised researchers, Cook said. But that strong signal also made sense, since greenhouse-gas emissions were rising during that time period. "From there, going into the mid-twentieth century, the signal disappears—probably due to aerosol pollution, masking the warming," Cook said. Industrial aerosols, which weren't controlled as pollutants until the 1970s, can affect regional cloud formation, rainfall, and temperature. Aerosols likely affected regional weather and also masked the impacts of climate change, even as greenhouse-gas emissions continued to rise. (In the spring of 2019, carbon dioxide levels in the atmosphere exceeded 410 parts per million.)

"In more recent decades, there has been the trend toward the climate-change signal we expect," Cook said. These trends are still imposed over regional patterns of wetting and drying that vary over time—patterns like El Niño and La Niña, which we experience here in the US Southwest. "Natural drought variability is important, we're not disputing that at all," he said. "But we're seeing changes, and we're seeing changes at these large scales that are not consistent with what we would expect from natural climate changes."

The 2010 United Nations IPCC assessment noted only "low confidence" in connecting the effects of climate change with the global hydroclimate. At the time, noted Cook, the body said it didn't have the tools or data to answer that question. "With this, we can say now that a near-global climate-change signal in regional drying and wetting trends is detectable," he said. "We can attribute these [trends] to climate change in the past, and we can expect that's going to continue out into the future."

Their study's fundamental question, Cook said, was, "Do the observations look like the models?" And the answer is, "Definitely yes." He explained that gives scientists more confidence to begin to nail down and understand how climate change is going to affect large-scale patterns of drying and wetting. He cautioned that the study looked at large-scale areas of the globe and does not hone in on specific regional drought events, like the Dust Bowl conditions in the United States, the drought of the 1950s in New Mexico, or the recent California drought.

"The word 'drought' can mean different things to different people, in different contexts," explained UNM's David Gutzler when I asked him to put the study into context for New Mexico. "The concept of drought is a little bit squishy and fuzzy."

In the *Nature* study, they looked at global-scale droughts. "These are droughts that they can identify from tree-ring records on different continents, all at the same time, and they can see basic, large-scale patterns of dryness that persist for a while," Gutzler said. "And what they find is a global coherence to it. But it's hard to map that concept onto a specific year and a specific place."

And, he said, the broad-scale propensity for drought goes up as the climate warms. As the study came out, New Mexico was coming off a bumper year for snowpack, and its rivers were rip-roaring. After the devastation of low water conditions the previous year, 2019 felt like a reprieve. Like we could put the lessons of bad years to work and implement plans that ensure we make it through bad years. And like it was okay to enjoy a green spring with high rivers.

Gutzler was among those who enjoyed the Rio Grande's high spring flows through Albuquerque in 2019. "My wife and I just took a walk near the Alameda Bridge [in Albuquerque] over the weekend, and it was wonderful," he said. "I think it's totally terrific to have an excellent snowpack and good runoff. But one good year doesn't remedy all the long-term water shortages we're dealing with."

———

There are reasons for hope, however. Not the "hope" of a campaign slogan, but hope that comes from a place of sleeplessness.

Elected in November 2018 to represent New Mexico's First Congressional District, Rep. Deb Haaland is among the first of two Native women elected to Congress. After her first week in Washington, DC, we'd agreed to meet at the Albuquerque BioPark's Botanic Garden to talk about climate change. And on a cold, cloudy morning, we ducked inside the garden's faux-cave, complete with giant toadstools and plaster footprints of prehistoric creatures. Neither warm nor particularly quiet, the cave is a uniquely terrible place to conduct an interview. Instead of being ruffled or appearing put out, Haaland laughs.

It's the kind of laugh that eliminates any speculation: what Haaland represents to the public has not eclipsed the person she is.

If you talk to people in the district, many were excited to cast their votes for a Native woman. A member of the Pueblo of Laguna, Haaland explained that tribes are "always the most underrepresented at any table." When asked how those proud to support a Native candidate can better support all tribal communities in New Mexico, she said Native people want what everyone wants—"a clean environment, a quality public education for their children, their elderly folks taken care of, health care for every citizen."

"Those are all the same things I am fighting for, for every single New Mexican," she said. "I think if we can just join forces, and we're strong allies together, we can always make sure that those things are possible for every single person in our state."

Throughout her campaign, Haaland was a vocal proponent of action on climate change. And when US House Speaker Nancy Pelosi announced in December 2018 that the newly Democratic-led House of Representatives would convene a Select Committee on the Climate Crisis, Haaland requested that committee assignment. "I realize that places like Florida, Louisiana, Houston, they will suffer because of the rising sea," she told me. "But climate change is going to affect the Southwest far more than so many other parts of the country."

In the past few years, increasingly urgent reports and models have shown how human-caused climate change is affecting the earth and will continue to exacerbate everything from sea-level rise and aridification to harm to the environment, public health, and the economy. But scientists have issued reports and warnings for decades. And yet, repeatedly over the years, when Democrats have held control over Congress, they have failed to act on climate change.

When asked what will be different this time, Haaland noted that a large number of the newly elected House Democrats campaigned specifically on climate change, and they are passionate about it. "The sheer number of us who are going to move the issue forward are present," she said, careful to note that action on climate change doesn't mean leaving behind workers in the fossil-fuel industry. "I know there are folks who have left the fossil-fuel industry to pursue other professions, and it works," she said. "What separates us from the animals is our ability to reason: Human beings always have opportunities to change gears and do something different."

And it's clear that humans need to do things differently.

"We need to stay on top of this," Haaland said. Many voices will continue advocating for fossil-fuel extraction, especially in a state like New Mexico that has been dependent upon oil and gas revenues for decades. But the signs of that dependency are everywhere, even in the sky—the largest methane anomaly, or "hotspot," in the northern hemisphere is above the Four Corners, she noted. "I think we need to put people's lives first, I think we need to put our environment first," she said. "Everything we have comes from our earth, and if we don't take care of it, we can expect to start losing things."

Pueblo people have grown food and nurtured crops in the high desert for centuries, she said. "It's always been hard, but we've done it." Then Haaland related a story about the Hopi Tribe, in Arizona's high desert. "They would have people who looked out—twenty-four hours a day, they would take shifts—and they would watch for when the water came down from the snowmelt," she said. "Because every year, after the snow finally melted and the water would come to their land, every member of the pueblo would come out and, with their implements, make sure that the water went down to their fields."

It was a community effort, their "one shot" at collecting water and ensuring it nourished their fields and crops. For pueblo people water has always been treated as precious. Her ancestors ensured she would have a future here today, she said, and she must have the same diligence, protecting land and water and keeping traditions alive. "It's worth losing sleep over, it's worth getting up in the middle of the night to nurture it so that life can be continued."

————

IN 2018 NEW Mexicans also elected a new governor—one who mentioned climate change and water challenges during her campaign, and who New Mexicans elected by a huge majority over a Republican congressman with deep ties to the oil and gas industry. Just weeks into the administration of Gov. Michelle Lujan Grisham, state agencies started moving forward on renewable energy, oil and gas facility inspections, climate-change initiatives, and trying to address the state's water challenges. Of course, Democrats have come and gone—and the climate has continued changing. It remains to be seen what can and will be done when it comes to curbing greenhouse-gas emissions and also making sure communities and ecosystems have a chance to adapt to the changes. There is serious talk about cutting methane emissions from the oil and gas industry, to follow the example of states like Colorado that have done so without federal regulations. But the rate at which drilling is predicted to continue, particularly in the Permian Basin, is worrisome.

As I mentioned earlier in the text, New Mexico became the fifth-largest oil-producing state in 2017, and it is also among the top natural gas–producing states. And according to the US Energy Information Administration, oil production in the Permian Basin, which includes New Mexico, is projected to continue increasing.

————

IN MAY 2019 middle and high school students around the state and across the country took over streets, plazas, and—in Albuquerque—a corner of the University of New Mexico's Johnson Field. Alyssa Ruiz from Sandia High School told the crowd that while the United States plans to spend more than a billion dollars building a wall along the US-Mexico border, the Trump administration's proposed budget for 2020 cuts spending on renewable energy. "When will our future be considered a national emergency?" she asked.

Katie Butler, a seventeen-year-old student from La Cueva High School, was among the co-organizers of the School Strike for Climate Action. "This isn't about politics," she said. Rather, climate change is a human-rights issue. And seventh grader Eliott Patton expressed the frustration echoed by many of the teens at the protest. "We need to get adults to stop Tweeting," she told the crowd, "and start acting."

The protest was part of a coordinated and global youth

movement on climate change, inspired at least in part by Swedish teen Greta Thunberg. And the protests continued, gaining momentum and numbers. In September 2019 Albuquerque youth activists successfully lobbied the city council to declare a climate emergency—and they organized another protest and walkout in solidarity with the Global Climate Strike. Speaking ahead of the protest, Patton, along with high schoolers Mariluz Lebkuechner and Olivia Gonzales, spoke with me about what's at stake for their generation.

"It's my future," fifteen-year old Gonzales said when I asked her why climate change—of all the challenges her generation faces—demanded her energy and attention. "When I was younger, I had my whole life planned out. Like, I knew when I was going to graduate college, I knew what I was going to do, I knew that I was going to have kids. Just the fact that's all in jeopardy is insane to me, that people still aren't listening to me. That's telling me [that] my future doesn't matter as much as yours."

Lebkuechner echoed that. "We all lie in bed at night, or daydream about our future," she said. "But if we don't do anything, that [future] might not happen."

The three also laid out their demands for New Mexico's political leaders: the first was to ensure the state is "100 percent renewable" by 2030. They also wanted the state's senators to sign on to the Green New Deal, sweeping federal legislation that would address climate change and social inequality, and they wanted a moratorium on hydraulic fracturing in the state. And, thirteen-year-old Patton added, "We want a 'just transition'—so that people aren't left behind. We use the money from the big oil and gas industries to create economic diversity for people who depend on the oil and gas industry."

The three said they wanted adults to ally with them—to strike with them and come to their events. But they need even more than that. "People told me that my generation needs to be the one to save the world," said Gonzales. "I never believed them. Or, if I did, I thought it would be when I was an adult, twenty-five or thirty.

But we don't have time left to do that. I'm fifteen, and I'm doing the most that I can for my age." She and her friends can't vote, she said. That can't pass bills. "That's really where we need the support of adults," she said. "To do what they can."

———

If it's worth being sleepless, it's also worth getting up in the middle of the night to look at the stars, and to spend days looking for birds and their nests. I remembered that while finishing this manuscript at Mi Casita, Aldo and Estella Leopold's cabin in Tres Peidras, New Mexico. In June 2019 I spent the month writing, wandering, stargazing, and reading a pile of books, including many of Leopold's that I hadn't read in years. "That land is a community is the basic concept of ecology," he wrote in 1948, in the foreword to *A Sand County Almanac*, "but that land is to be loved and respected is an extension of ethics. That land yields a cultural harvest is a fact long known, but latterly forgotten."

It's fair to assume that Leopold would have been appalled by our inaction on climate change—especially since even in the 1940s he was lamenting that "our bigger-and-better society is now like a hypochondriac, so obsessed with its own economic health as to have lost the capacity to remain healthy."

In between reading and writing—or, oftentimes with a book and notebook tucked in a backpack alongside a camera, binoculars, and a water bottle—I explored the Carson National Forest, just outside the cabin door. Seeking the perfect sunset-viewing spot my first evening at Mi Casita, I scrambled atop the rock formation behind the cabin. Clambering west and scaling a pile of rocks, I was met by two angry ravens. They swooped and yelled, and I got the message: There was a nest nearby, and I needed to get lost. I staked out a different viewing spot, watched the sun sink behind purple clouds, and got a good night's sleep. The next morning I awoke, wrote, and then set out to find the nest, but from a spot that wouldn't stress out the birds. After finding the nest—tucked onto a cliff shelf and made of sticks, scat, baling twine, and a dead rodent—I spent weeks watching the chicks. Initially the four sat silently in their nest, their blue eyes milky. Then they learned to sit with their beaks open. After a little while longer, they began silently exploring the cliff face. The parents still kept an eye on me, landing atop a nearby ponderosa pine or flying low overhead to let me know I wasn't fooling anyone with my hiding spot. In between their near-constant tasks—feeding the four chicks, plus chasing off raptors, other ravens, and a committee of turkey vultures who would soar over every afternoon—the two adults would sometimes sit within the branches of a pine, ignore me, and groom one another.

Toward the end of June I was tucked into a divot in the granite after dinner. My dog, a watchful red heeler who hates the rain, camped out beside me while I scrawled in a notebook and occasionally peeked at the young ravens through binoculars. As sunset approached, the wind picked up and the skies spit a light rain. The dog was antsy, and I'd grown cold. Even while the rest of the state bakes in June, in northern New Mexico it's a good idea to have a coat if you're planning to linger outside past dark or in the rain. Before packing up and heading back to the cabin, I took one last look at the crag. Instead of having returned to the nest, though, all four chicks had spread across the cliff face. As the wind picked up, they flapped their wings, five-six-seven-eight beats in a row. A couple were hopping. And the parents were watching, silently, from two ponderosas between the nest and me. It was then, with the sun going down and a storm barreling in from the west, that the four ravens decided to fledge. All at once they descended from the cliff face, each pointed in a slightly different direction. They spread their wings. And took a chance.

We can do the same.

Notes

Preface

In the preface I draw upon a 2015 memo from the New Mexico Legislative Finance Committee, which provides analysis for legislatures. In that memo, analysts explained that a $1 change in the price of oil equals about a $10 million change in the amount of cash in the state's general fund. They also wrote that "sensitivity to oil has dramatically increased in recent years, magnified due to the record levels of production." While that number can change over time, and the oil and gas industry pegs that sensitivity to about a $13 million change in the general fund, I chose to go with the LFC's numbers, since that committee is nonpartisan and its role is to provide unbiased analysis.

Introduction

In the fall of 2018—two years after the election of President Donald Trump—the reports on climate change were becoming increasingly urgent. In addition, scientists were starting to more definitively make connections between climate change and certain extreme weather events. There are a number of reports readers can look at in more detail, including the US Global Change Program's Fourth National Climate Assessment, which is a companion to a volume published in 2017 regarding the physical climate science, and the IPCC's special report on how a 2°C rise in global temperature will affect the planet, its ecosystems, and its human communities, compared with a 1.5°C temperature increase. These can be found online at http://nca2018.globalchange.gov and http://ipcc.ch/report/sr15/.

The two Colorado River studies mentioned in the introduction include Udall and Overpeck, "The Twenty-First Century Colorado River"; and Xiao, Udall, and Lettenmaier, "On the Causes of Declining Colorado River Streamflows."

To read more about drought in the southwestern tree-ring record, see Woodhouse et al., "A 1,200-year perspective of 21st century drought"; Cook, Ault, and Smerdon, "Unprecedented 21st century drought risk; Cook et al., "Megadroughts in North America"; Meko et al., "Medieval drought in the Upper Colorado River Basin"; and Routson, Woodhouse, and Overpeck, "Second century megadrought in the Rio Grande headwaters." And two resources on birds and climate change include Both et al., "Climate change and population declines "; and Audubon's Birds and Climate Change Report, http://climate.audubon.org/, accessed October 21, 2017.

Chapter 1

In the spring of 2011 I was juggling raising a young daughter and trying to continue making a living as a journalist, now as a freelancer. Parts of this chapter first appeared as feature stories in the *Santa Fe Reporter*, the City Different's scrappy alt-weekly whose editors (Julia Goldberg and then, later, Julie Ann Grimm) consistently ran stories about climate change at a time when the mainstream media ignored or obfuscated the issues. As many New Mexicans will remember, the spring and summer of 2011 was a smoky one. There were fires all around the Southwest, including the Wallow Fire in Arizona, whose smoke inundated Albuquerque, and Las Conchas, which changed the way many of us look at summer thunderheads. The lecture of Eric Blinman's that had such an impact on me was titled "The Rear View Window: 2,000 Years of People and Climate Change in the Southwest," and he delivered it in Albuquerque, New Mexico, in 2010.

The Horseshoe Two Fire in southern Arizona ignited the morning of May 8, 2011. The human-caused fire burned 9,000 acres on its first day, and 222,954 acres in total. The Wallow Fire started in Arizona on May 29, 2011, and spread into New Mexico, eventually burning 538,049 acres, mostly in Arizona. And Las Conchas started on the afternoon of June 26, 2011.

Chapter 2

It's tempting to ignore politics when writing about climate change and the environment. No matter how stark a study or burned out a forest, it's far more delightful to jump down the rabbit hole of scientific data and methods or tromp around a mountainside. Covering politics, whether international, national, statewide, or local, is not as much fun. Unfortunately, though, it's important. No matter how much scientists understand and communicate about the reasons behind climate change and the impacts warming will have on ecosystems and human communities, action on climate change is left largely in the hands of politicians. This became all too clear to me, especially once I attended the Parties of the United Nations Framework Convention on Climate Change in Cancún, Mexico, in 2010, thanks to support from the Earth Journalism Network, and then came home to report on the administration of Gov. Susana Martinez. And then again in 2017, when I began reporting on the administration of President Donald Trump.

During the Bush administration I reported extensively on political interference at federal agencies for *High Country News*. Some of those stories, which are also available online, include: "Are minnow scientists under the gun?" (June 23, 2003), "Sounds science goes sour," (June 23, 2003), "Bush undermines bedrock environmental law" (October 28, 2002), "Stand your ground" (December 20, 2004), and "Conscientious objectors" (Dec. 20, 2004).

The Stewart Udall essay, "A message to our grandchildren," appeared in multiple media outlets, including in *High Country News* on March 31, 2008.

Chapter 3

Beginning in the fall of 2014, and for much of 2015, I reported for

KUNM-FM on oil and gas issues in northwestern New Mexico. All the reporting from that series is online at https://www.kunm.org/term/drilling-deep.

Portions of this chapter also come from stories for the *Santa Fe Reporter* and *New Mexico Political Report*, including in 2017, when I interviewed State Reps. Derrick Lente and Daniel Tso. Since that time the fight over drilling in northwestern New Mexico, and around Chaco Canyon, has gained increased attention from the state's congressional delegation, especially Sens. Tom Udall and Martin Heinrich and Reps. Deb Haaland and Ben Ray Luján.

Chapter 4

As the impacts of climate change became more evident, and the twenty-first-century drought dragged on, it became clear to me that these changes—and writing about them on a regular basis—exact an emotional toll. And it's not only farmers and ranchers and other people whose livelihoods depend directly on the land and water who feel the strain of drought, wildfire, and forest die-offs. The more we understand about climate change and our role in changing the planet, the harder it is to ignore the fact that as a culture, Americans don't have a clue how to mourn these landscapes or come to terms with how our lifestyles have wreaked havoc on the planet and its future.

Parts of this chapter appeared as an essay in the *Santa Fe Reporter*, and the response I received after sharing my own emotional struggles was interesting. A few people were angry. At least one person told me it was unhealthy to write that there's no time for mourning or grief—and disrespectful to people who are mourning or grieving. But most people who contacted me seemed to be grateful for the attempt at figuring out how to feel about the loss of forests and species right here in our own state. I appreciate Larry Rasmussen's gentle thoughtfulness and his invitation to meet with him at his and his wife's home in 2015. I'm also grateful to Christopher Witt, with the Museum of Southwestern

Biology at the University of New Mexico, who likely got more than he bargained for when I asked to visit the collection of birds the museum houses. For stories about the minnow, Christopher Hoagstrom endured interviews in 2002, 2003, 2015, and 2016. And he furnished me with one of my all-time favorite quotes in almost two decades of reporting. In 2002, while he was still working for the Fish and Wildlife Service, a frustrated Chris Hoagstrom told me, "If society consciously decided, 'We don't give a shit about little fish'—if we put it on the books and said, 'We, as Americans, decide little fish don't mean a thing to us,' then, whatever, that's what we've decided. But instead, we've listed these fish, we've got the Endangered Species Act, and we just don't apply it."

The studies on forest mortality include Allen and Breshears, "Drought-induced shift"; Allen, "Interactions across spatial scales"; Allen et al., "A global overview of drought"; Mcdowell et al., "Multi-scale predictions of massive conifer mortality"; and Williams et al., "Temperature as a potent driver."

Chapter 5

If you haven't read Pope Francis's encyclical letter, "On Care for Our Common Home, Laudato Si," I can't recommend it highly enough, even if you're not Catholic or religious in any way. It's available as a printed book, and also online at https://laudatosi. com/watch. Rasmussen's book, *Earth-Honoring Faith*, will also be useful to those who are trying to navigate their spirituality with environmentalism. The interviews in this chapter, with Rasmussen, Joan Brown, Javier Benavidez, Father Frank Quintana, and Patricia Gallegos, all took place in 2015.

Chapter 6

Between Election Day 2016 and the inauguration of President Donald Trump, a scientist I'd never spoken with before invited me to visit the tree-ring laboratory in Santa Fe and learn about the work he and a colleague do on forest planning and wildfire.

As our meeting approached—and headlines continued flaring up about scientists being muzzled by their agencies—I expected him to cancel.

Halfheartedly I checked in, dropping him an email: "Does the 31st still work for you guys?"

A few hours later he pinged a response: "Yep. 1pm right?"

Phew. Scientists. Talking with me about their work.

When Collin Haffey and Ellis Margolis met me in the foyer of their office in Santa Fe, the first thing I noticed was their smiles. These weren't beaten-down scientists cowering before the administration's hammer on the agency. They were two guys in jeans, one with mud on his boots, who wanted to talk about forests and tree rings, climate change and wildfires.

Haffey took me out to the burn scar of Las Conchas twice, acting as a tour guide to both hope and despair—but mostly to the science of ecology and land management. Over the course of years he's answered piles of my questions, and though he no longer works at USGS, I'm heartened by the work that he and many of his younger colleagues do on climate change.

To read some of the nitty gritty details about the 2011 Las Conchas fire, there's the Southwest Fire Science Consortium's fact sheet on the fire: http://swfireconsortium.org/wp-content/uploads/2014/12/Las-Conchas-Factsheet.pdf; and the National Park Service's document: https://www.nps.gov/band/learn/nature/lasconchas.htm. It's in that NPS document that I learned "over 75% of Frijoles Canyon lay within the fire's footprint, much of it burned with high severity."

I learned even more about the fire and its impacts from the National Park Service's Robert Parmenter, who's lucky enough to work at the Valles Caldera National Wildlife Refuge, and Jeremy Sweat, at Bandelier National Monument. I also learned a great deal from the various participants of the Southwest Jemez Mountains Restoration Project Annual All-Hands Meeting in Santa Fe, New Mexico, in April 2017.

Chapter 7

This chapter came together from various reporting trips to the Sandia and Jemez Mountains, including to the home and tree-ring lab of Tom Swetnam in 2017 and 2018. Parts of this chapter appeared in *New Mexico Political Report*, and also within a monthly television program on KNME-TV's *New Mexico In Focus*. The film crew for "Our Land: New Mexico's Environmental Past, Present, and Future" at this time included Antony Lostetter, Sarah Gustavus, and Antony Rodriguez. A special thanks to Kerry Jones, who in this chapter talks about the dry winter of 2017/2018 but has also been a longtime source and someone who is always willing to talk with reporters and is able to explain weather and climate change to the public in ways that are digestible. "Climate is what you expect," I once heard Jones tell a classroom full of teachers who were learning about climate change and tree rings. "Weather is what you get." Climate represents general conditions in a particular region—thirty years is the standard time period experts use for talking about it, he explained. Meanwhile, weather happens on a day-to-day basis, and it is the fluctuating state of the atmosphere characterized by temperature, precipitation, wind, clouds, and other elements. Think about it another way, Jones told the teachers: Climate trains the boxer, while weather throws the punches.

Chapter 8

An update to this chapter: The permit to drill groundwater from the San Augustin Plains was rejected again in 2019. But the company is appealing again.

Peggy Johnson spoke at the New Mexico Water Dialogue meeting in Albuquerque on January 8, 2015. Teri Farley and Garrett Petrie were kind enough to allow us to film at their home (and meet their cute animals) in 2017 for our *New Mexico In Focus* show on NM PBS, "Our Land: New Mexico's Environmental Past, Present, and Future." Interviews with Anita Hand, Rep. Gail

Armstrong, Douglas Meiklejohn, and Carol Pittman took place over the phone in 2017. Michel Jichlinski, Whitney Waite, and Melissa Dosher, with the Office of the State Engineer, corresponded via email in 2017, and that same year I spoke with Pete and Eva Dempsey and Joan Brooks in person, following the San Augustin Plains meeting in Socorro.

Studies cited in this chapter include Meixner et al., "Implications of project climate change"; Niraula et al., "How might recharge change"; Perrone and Jasechko, "Groundwater wells in the western United States"; and Rawling and Rinehart, "Lifetime projections for the High Plains Aquifer." To read one of the definitive books on how humans altered the southwestern desert through the use of fossil fuels and groundwater pumping, pick up a copy of Charles Bowden's 1977 book, *Killing the Hidden Waters*.

Chapter Nine

The various studies cited in this chapter include Ajami et al., "Seasonalizing Mountain System Recharge"; Chavarria and Gutzler, "Observed changes in climate and streamflow"; Cook, Ault, and Smerdon, "Unprecedented 21st century drought risk"; Dettinger, Udall, and Georgakakos, "Western water and climate change"; Elias et al., "Climate change, agriculture and water resources"; Elias et al., "Assessing climate change impacts"; Gutzler and Robbins, "Climate variability and projected change"; Lehner et al., "Assessing recent declines in Upper Rio Grande"; Mankin et al., "The potential for snow"; Prein et al., "Running dry"; Ruhí, Olden, and Sabo, "Declining streamflow induces collapse"; Seager et al., "Model projections of an imminent transition"; Udall and Overpeck, "The twenty–first century Colorado River hot drought"; Ward and Booker, "Economic costs and benefits of instream flow protection"; and Wuebbles, "Climate science special report."

Chapter 10

This chapter was reported on the ground in the fall of 2018 in southern New Mexico. I've spent years reporting on the US

Supreme Court lawsuit, Texas v. New Mexico & Colorado, and gained valuable insight into the lawsuit and the parties' positions and roles when *NM Political Report*, the *Santa Fe Reporter*, and KNME-TV all pooled their resources and sent me to Washington, DC, to report on oral arguments in January 2018.

Afterword

The study cited in this chapter is Marvel et al., "Twentieth-century hydroclimate changes."

Bibliography

Ajami, Hoori, Thomas Meixner, Francina Dominguez, James Hogan, and Thomas Maddock III. "Seasonalizing Mountain System Recharge in Semi-Arid Basins-Climate Change Impacts." *Groundwater* 50, no. 4 (2012): 585–97.

Allen, Craig D. "Interactions across spatial scales among forest dieback, fire, and erosion in northern New Mexico landscapes." *Ecosystems* 10, no. 5 (2007): 797–808.

Allen, Craig D., and David D. Breshears. "Drought-induced shift of a forest–woodland ecotone: rapid landscape response to climate variation." *Proceedings of the National Academy of Sciences* 95, no. 25 (1998): 14839–14842.

Allen, Craig D., Alison K. Macalady, Haroun Chenchouni, Dominique Bachelet, Nate McDowell, Michel Vennetier, Thomas Kitzberger et al. "A global overview of drought and heat-induced tree mortality reveals emerging climate change risks for forests." *Forest ecology and management* 259, no. 4 (2010): 660–84.

Audubon Society. Birds and Climate Change Report, http://climate.audubon.org, accessed October 21, 2017.

Austin, Mary. *The Land of Little Rain*. Albuquerque: University of New Mexico Press, 1974.

Backus, George A. *Assessing the near-term risk of climate uncertainty: interdependencies among the US states*. No. SAND2011-7846C. Sandia National Laboratories (SNL-NM), Albuquerque, NM (United States), 2011.

Betancourt, Julio L., Elizabeth A. Pierson, K. Aasen Rylander, James A. Fairchild-Parks, and Jeffrey S. Dean. "Influence of history and climate on New Mexico pinon-juniper woodlands." *Managing pinon-juniper ecosystems for sustainability and social needs. Gen. Tech. Rep. RM-236. Fort Collins, CO: US Department of Agriculture, Forest Service, Rocky Mountain Forest and Range Experiment Station* (1993): 42–62.

Booker, James F., and Frank A. Ward. "Instream flows and endangered

species in an international river basin: The Upper Rio Grande."
American journal of agricultural economics 81, no. 5 (1999):
1262–1267.

Both, Christiaan, Sandra Bouwhuis, C. M. Lessells, and Marcel E. Visser.
"Climate change and population declines in a long-distance migra-
tory bird." *Nature* 441, no. 7089 (2006): 81–83.

Both, Christiaan, Chris AM Van Turnhout, Rob G. Bijlsma, Henk Sie-
pel, Arco J. Van Strien, and Ruud PB Foppen. "Avian population
consequences of climate change are most severe for long-distance
migrants in seasonal habitats." *Proceedings of the Royal Society of
London B: Biological Sciences* (2009): rspb20091525.

Both, Christiaan, and Marcel E. Visser. "Adjustment to climate change
is constrained by arrival date in a long-distance migrant bird."
Nature 411, no. 6835 (2001): 296–98.

Bowden, Charles. *Killing the Hidden Waters*. Austin: University of Texas
Press, 1977.

Chavarria, Shaleene B., and David S. Gutzler. "Observed changes in
climate and streamflow in the upper Rio Grande Basin." *JAWRA:
Journal of the American Water Resources Association* 54, no. 3
(2018): 644–59.

Clark, Ira G. *Water in New Mexico*. Albuquerque: University of New
Mexico Press, 1987.

Cook, Benjamin I. *Drought: An Interdisciplinary Perspective*. New York:
Columbia University Press, 2019.

Cook, Benjamin I., Toby R. Ault, and Jason E. Smerdon. "Unprecedented
21st century drought risk in the American Southwest and Central
Plains." *Science Advances* 1, no. 1 (2015): e1400082.

Cook, Edward R., Richard Seager, Richard R. Heim, Russell S. Vose,
Celine Herweijer, and Connie Woodhouse. "Megadroughts in
North America: Placing IPCC projections of hydroclimatic change
in a long-term palaeoclimate context." *Journal of Quaternary Science* 25, no. 1 (2010): 48–61.

Damage Assessment and Current Actions: Santa Clara Pueblo Canyon.
Report to the New Mexico Legislative Health and Human Services
Committee, 2014.

Dettinger, Michael, Bradley Udall, and Aris Georgakakos. "Western
water and climate change." *Ecological Applications* 25, no. 8
(2015): 2069–2093.

Doerr, Anthony. "The Sword of Damocles: On suspense, shower mur-
ders, and shooting people on the beach," *In The Writer's Notebook
II: Craft Essays from Tin House* (2012): 57–82.

Elias, Emile, Al Rango, Ryann Smith, Connie Maxwell, Caiti Steele, and

Kris Havstad. "Climate change, agriculture and water resources in the Southwestern United States." *Journal of Contemporary Water Research & Education* 158, no. 1 (2016): 46–61.

Elias, E. H., A. Rango, C. M. Steele, J. F. Mejia, and R. Smith. "Assessing climate change impacts on water availability of snowmelt-dominated basins of the Upper Rio Grande basin." *Journal of Hydrology: Regional Studies* 3 (2015): 525–46.

Feng, Shuaizhang, Alan B. Krueger, and Michael Oppenheimer. "Linkages among climate change, crop yields and Mexico–US cross-border migration." *Proceedings of the National Academy of Sciences* 107, no. 32 (2010): 14257–62.

Flato, Gregory, Jochem Marotzke, Babatunde Abiodun, Pascale Braconnot, Sin Chan Chou, William J. Collins, Peter Cox et al. "Evaluation of Climate Models. In: Climate Change 2013: The Physical Science Basis. Contribution of Working Group I to the Fifth Assessment Report of the Intergovernmental Panel on Climate Change." *Climate Change 2013* 5 (2013): 741–866.

Frankenberg, Christian, Andrew K. Thorpe, David R. Thompson, Glynn Hulley, Eric Adam Kort, Nick Vance, Jakob Borchardt et al. "Airborne methane remote measurements reveal heavy-tail flux distribution in Four Corners region." *Proceedings of the National Academy of Sciences* (2016): 201605617.

Garfin, Gregg, Guido Franco, Hilda Blanco, Andrew Comrie, Patrick Gonzalez, Thomas Piechota, Rebecca Smyth, and Reagan Waskom. "Southwest: The Third National Climate Assessment." In *Climate change impacts in the United States: The third national climate assessment*. US Global Change Research Program, 2014.

Gori, David, Martha S. Sooper, Ellen S. Soles, Mark Stone, Ryan Morrison, Thomas F. Turner, David L. Propst, Gregg Grafin, and Kelly Kindscher. "Gila River flow needs assessment." (2016).

Griffin, Daniel, Connie A. Woodhouse, David M. Meko, David W. Stahle, Holly L. Faulstich, Carlos Carrillo, Ramzi Touchan, Christopher L. Castro, and Steven W. Leavitt. "North American monsoon precipitation reconstructed from tree-ring latewood." *Geophysical Research Letters* 40, no. 5 (2013): 954–58.

Grover, Herbert D., and H. Brad Musick. "Shrubland encroachment in southern New Mexico, USA: an analysis of desertification processes in the American Southwest." *Climatic Change* 17, no. 2 (1990): 305–30.

Gutzler, David S. "Climate and drought in New Mexico." *Water Policy in New Mexico: Addressing the Challenge of an Uncertain Future* (2013): 56.

————. "Streamflow Projections for the Upper Gila River." *New Mexico Interstate Stream Commission* (2013).

Gutzler, David S., Deirdre M. Kann, and Casey Thornbrugh. "Modulation of ENSO-based long-lead outlooks of southwestern US winter precipitation by the Pacific decadal oscillation." *Weather and Forecasting* 17, no. 6 (2002): 1163–1172.

Gutzler, David S., and Tessia O. Robbins. "Climate variability and projected change in the western United States: regional downscaling and drought statistics." *Climate Dynamics* 37, no. 5–6 (2011): 835–49.

Hand, J. L., W. H. White, K. A. Gebhart, N. P. Hyslop, T. E. Gill, and B. A. Schichtel. "Earlier onset of the spring fine dust season in the southwestern United States." *Geophysical Research Letters* 43, no. 8 (2016): 4001–4009.

Horgan, Paul. *Great River*. New York: Rinehart and Company, 1954.

Intergovernmental Panel on Climate Change. *Climate Change 2014– Impacts, Adaptation and Vulnerability: Regional Aspects*. Cambridge University Press, 2014.

Intergovernmental Panel on Climate Change. *Global warming of 1.5°C. An IPCC Special Report on the impacts of global warming of 1.5°C above pre-industrial levels and related global greenhouse gas emission pathways, in the context of strengthening the global response to the threat of climate change, sustainable development, and efforts to eradicate poverty*. Eds. V. Masson-Delmotte, P. Zhai, H. O. Pörtner, D. Roberts, J. Skea, P. R. Shukla, A. Pirani, W. Moufouma-Okia, C. Péan, R. Pidcock, S. Connors, J. B. R. Matthews, Y. Chen, X. Zhou, M. I. Gomis, E. Lonnoy, T. Maycock, M. Tignor, T. Waterfield, 2018.

"Las Conchas Fire, Jemez Mountains," fact sheet, Southwest Fire Science Consortium, n.d. http://swfireconsortium.org/wp-content/uploads/2014/12/Las-Conchas-Factsheet.pdf.

Lehner, Flavio, Eugene R. Wahl, Andrew W. Wood, Douglas B. Blatchford, and Dagmar Llewellyn. "Assessing recent declines in Upper Rio Grande runoff efficiency from a paleoclimate perspective." *Geophysical Research Letters* 44, no. 9 (2017): 4124–4133.

Jardine, Angela, Robert Merideth, Mary Black, and Sarah LeRoy. *Assessment of climate change in the southwest United States: a report prepared for the National Climate Assessment*. Island Press, 2013.

Kennedy, Thomas L., David S. Gutzler, and Ruby L. Leung. "Predicting future threats to the long-term survival of Gila trout using a high-resolution simulation of climate change." *Climatic Change* 94, no. 3 (2009): 503–15.

Kort, Eric A., Christian Frankenberg, Keeley R. Costigan, Rodica Lindenmaier, Manvendra K. Dubey, and Debra Wunch. "Four corners: The largest US methane anomaly viewed from space." *Geophysical Research Letters* 41, no. 19 (2014): 6898–6903.

"The Las Conchas Fire." US Department of the Interior, National Park Service, n.d. https://www.nps.gov/band/learn/nature/lasconchas.htm.

Llewellyn, Dagmar, and Seshu Vaddey. West-wide climate risk assessment: Upper Rio Grande impact assessment. Albuquerque: US Bureau of Reclamation, Albuquerque Area Office, Upper Colorado Region. 2013.

Mankin, Justin S., Daniel Viviroli, Deepti Singh, Arjen Y. Hoekstra, and Noah S. Diffenbaugh. "The potential for snow to supply human water demand in the present and future." *Environmental Research Letters* 10, no. 11 (2015): 114016.

Marlon, Jennifer, Peter Howe, Matto Mildenberger, and Anthony Leiserowitz. "Yale Climate Opinion Maps—U.S. 2016." Yale Program on Climate Change Communication. http://climatecommunication.yale.edu/visualizations-data/ycom-us-2016/?est=happening&-type=value&geo=county.

Marvel, Kate, Benjamin I. Cook, Céline J. W. Bonfils, Paul J. Durack, Jason E. Smerdon, and A. Park Willliams. "Twentieth-century hydroclimate changes consistent with human influence." *Nature* 569 (2019): 59–72.

Mcdowell, Nathan Gabriel, A. P. Williams, Chonggang Xu, W. T. Pockman, L. T. Dickman, Sanna Sevanto, R. Pangle et al. "Multi-scale predictions of massive conifer mortality due to chronic temperature rise." *Nature Climate Change* 6, no. 3 (2016): 295.

Meixner, Thomas, Andrew H. Manning, David A. Stonestrom, Diana M. Allen, Hoori Ajami, Kyle W. Blasch, Andrea E. Brookfield, Christopher L. Castro, Jordan F. Clark, David J. Gochis, Alan L. Flint, Kristin L. Neff, Rewati Niraula, Matthew Rodell, Bridget R. Scanlon, Kamini Singha, and Michelle A. Walvoord. "Implications of project climate change for groundwater recharge in the western United States," *Journal of Hydrology* 534 (March 2016): 124–38.

Meko, David M., Connie A. Woodhouse, Christopher A. Baisan, Troy Knight, Jeffrey J. Lukas, Malcolm K. Hughes, and Matthew W. Salzer. "Medieval drought in the upper Colorado River Basin." *Geophysical Research Letters* 34, no. 10 (2007).

Meyer, J. L., M. J. Sale, P. J. Mulholland, and N. L. Poff. "Impacts of climate change on aquatic ecosystem functioning and health." *JAWRA: Journal of the American Water Resources Association*, 35, no. 6 (1999): 1373–1386.

Moritz, Max A., Jon E. Keeley, Edward A. Johnson, and Andrew A. Schaffner. "Testing a basic assumption of shrubland fire management: how important is fuel age?." *Frontiers in Ecology and the Environment* 2, no. 2 (2004): 67–72.

Muller, Edward K. (Ed.) *DeVoto's West*. Athens: Swallow Press, 2005.

Niraula, R., T. Meixner, F. Dominguez, N. Bhattarai, M. Rodell, H. Ajami, D. Gochis, and C. Castro. "How might recharge change under projected climate change in the western US?." *Geophysical Research Letters* (2017).

Pascolini-Campbell, M. A., Richard Seager, David S. Gutzler, Benjamin I. Cook, and Daniel Griffin. "Causes of interannual to decadal variability of Gila River streamflow over the past century." *Journal of Hydrology: Regional Studies* 3 (2015): 494–508.

Perrone, D. and S. Jasechko. "Groundwater wells in the western United States." *Environmental Research Letters* 12 (2017): 1–10.

Pope Francis. *On Care for Our Common Home: Laudato Sí*. Washington, DC: United States Conference of Catholic Bishops, 2015.

Popovich, Nadja, John Schwartz, and Tatiana Schlossberg. "How Americans think about climate change, in six maps." *The New York Times*, March 21, 2017.

Prein, Andreas F., Gregory J. Holland, Roy M. Rasmussen, Martyn P. Clark, and Mari R. Tye. "Running dry: The US Southwest's drift into a drier climate state." *Geophysical Research Letters* 43, no. 3 (2016): 1272–1279.

President's Science Advisory Committee—Environmental Pollution Panel. "Restoring the Quality of Our Environment." November 1965.

Price, V. B. *The Orphaned Land*. Albuquerque: University of New Mexico Press, 2011.

Rasmussen, Larry L. *Earth-Honoring Faith*. Oxford: University Press, 2013.

Rawling, Geoffrey C., and Alex J. Rinehart. "Lifetime projections for the High Plains Aquifer in east-central New Mexico." New Mexico Bureau of Geology and Mineral Resources, Open-File Report 591, July 2017.

Reclamation (Bureau of Reclamation). *SECURE Water Act Section 9503 (c)—Reclamation Climate Change and Water*. Denver, CO: Bureau of Reclamation, Policy and Administration, 2016.

Routson, Cody C., Connie A. Woodhouse, and Jonathan T. Overpeck. "Second century megadrought in the Rio Grande headwaters, Colorado: How unusual was medieval drought?" *Geophysical Research Letters* 38, no. 22 (2011).

Ruhí, Albert, Julian D. Olden, and John L. Sabo. "Declining streamflow induces collapse and replacement of native fish in the American Southwest." *Frontiers in Ecology and the Environment* 14, no. 9 (2016): 465–72.

Santos, Fernanda. *The Fire Line*. New York: Flatiron Books, 2016.

Seager, Richard, Mingfang Ting, Isaac Held, Yochanan Kushnir, Jian Lu, Gabriel Vecchi, Huei-Ping Huang et al. "Model projections of an imminent transition to a more arid climate in southwestern North America." *Science* 316, no. 5828 (2007): 1181–1184.

Scurlock, Dan. *From the Rio to the Sierra: An Environmental History of the Middle Rio Grande Basin*. General Technical Report RMRS-GTR-5. Fort Collins, CO: United States Department of Agriculture, Forest Service, 1998.

Shabecoff, Philip. "Global Warming Has Begun, Expert Tells Senate." *New York Times*, June 24, 1988.

Stahle, David W., Edward R. Cook, Malcolm K. Cleaveland, Matthew D. Therrell, David M. Meko, Henri D. Grissino-Mayer, Emma Watson, and Brian H. Luckman. "Tree-ring data document 16th century megadrought over North America." *Eos, Transactions American Geophysical Union* 81, no. 12 (2000): 121–25.

Stanton, Jennifer S., Sharon L. Qi, Derek W. Ryter, Sarah E. Falk, Natalie A. Houston, Steven M. Peterson, Stephen M. Westenbroek, and Scott C. Christenson. "Selected approaches to estimate water-budget components of the High Plains, 1940 through 1949 and 2000 through 2009." *US Geol. Surv. Sci. Invest. Rep* 5183 (2011): 79.

State of New Mexico Agency Technical Working Group. "Potential Effects of Climate Change on New Mexico." December 30, 2005. https://www.env.nm.gov/aqb/cc/Potential_Effects_Climate_Change_NM.pdf

Stuart, David E. *Anasazi America*. Albuquerque: University of New Mexico Press, 2000.

Tong, Daniel Q., Julian XL Wang, Thomas E. Gill, Hang Lei, and Binyu Wang. "Intensified dust storm activity and Valley fever infection in the southwestern United States." *Geophysical Research Letters* 44, no. 9 (2017): 4304–4312.

Udall, B., & Overpeck, J. "The twenty-first century Colorado River hot drought and implications for the future." *Water Resources Research* 53, no. 3 (2017): 2404–2418.

USGCRP. *Impacts, Risks, and Adaptation in the United States: Fourth National Climate Assessment, Volume II*. Ed. D. R. Reidmiller, C. W. Avery, D. R. Easterling, K. E. Kunkel, K. L. M. Lewis, T. K. Maycock, and B. C. Stewart. US Global Change Research Program, Washington, DC, 2018. doi: 10.7930/NCA4.2018.

Visser, Marcel E., and Christiaan Both. "Shifts in phenology due to global climate change: the need for a yardstick." *Proceedings of the Royal Society of London B: Biological Sciences* 272, no. 1581 (2005): 2561–2569.

Visser, Marcel E., Christiaan Both, and Marcel M. Lambrechts. "Global climate change leads to mistimed avian reproduction." *Advances in ecological research* 35 (2004): 89–110.

Ward, Frank A., and James F. Booker. "Economic costs and benefits of instream flow protection for endangered species in an international basin 1." *JAWRA: Journal of the American Water Resources Association* 39, no. 2 (2003): 427–40.

Weiss, Jeremy L., David S. Gutzler, Julia E. Allred Coonrod, and Clifford N. Dahm. "Seasonal and inter-annual relationships between vegetation and climate in central New Mexico, USA." *Journal of Arid Environments* 57, no. 4 (2004): 507–34.

Westerling, Anthony L., Hugo G. Hidalgo, Daniel R. Cayan, and Thomas W. Swetnam. "Warming and earlier spring increase western U.S. forest wildfire activity." *Science* 313, no. 5789 (2006): 940–43.

Williams, A. Park, Craig D. Allen, Alison K. Macalady, Daniel Griffin, Connie A. Woodhouse, David M. Meko, Thomas W. Swetnam et al. "Temperature as a potent driver of regional forest drought stress and tree mortality." *Nature climate change* 3, no. 3 (2013): 292–97.

Woodhouse, Connie A., David M. Meko, Glen M. MacDonald, Dave W. Stahle, and Edward R. Cook. "A 1,200-year perspective of 21st century drought in southwestern North America." *Proceedings of the National Academy of Sciences* 107, no. 50 (2010): 21283–1288.

Woodhouse, Connie A., and Jonathan T. Overpeck. "2000 years of drought variability in the central United States." *Bulletin of the American Meteorological Society* 79, no. 12 (1998): 2693–2714.

Wuebbles, Donald J., David W. Fahey, and Kathy A. Hibbard. "Climate science special report: fourth national climate assessment, volume I." 2017.

Xiao, Mu, Bradley Udall, and Dennis P. Lettenmaier. "On the causes of declining Colorado River streamflows." *Water Resources Research*, 54 (2018): 6739–6756.

Index

Page numbers in italic text indicate illustrations.

Nez, Jonathan, 58

Obama, Barack: administration of, 27, 28, 29, 30, 31, 32, 56, 81
Office of the State Engineer, NM, 118, 123, 124–125, 126–127
Ogallala, 115
oil, industry and production in New Mexico: xiii, xiv, 5, 28, 39, 40, 43–44, 49–50, 51, 52–57, 59–62
"On Care for Our Common Home, Laudato Si,'" 75–76, 77–81
Oso Complex Fire, 97
Overpeck, Jonathan, 8–9, 11–12, 84–85, 136–37, 166, 172

Pachuauri, Rajendra, 31–32
Perrone, Debra, 114–16
Paris Accord (Paris Agreement), 34–36, 38, 82, 84–85
Parmenter, Bob, 107–8
Pecos River, 40, 70
Pelz, Jen, 148–49
Permian Basin, xiii–xiv, 60, 161
Plains of San Agustin, 113, 122–27
Pope Francis, 75–76, 77–81
pyrocumulous cloud, 17
Powell, Lewis, 36–37
Powell Manifesto, 36–37

Rasmussen, Larry, 65–66, 75–76, 168, 169
Republic of Seychelles, 22
Richardson, Bill, 39
Rio Grande, xi, 11, 29, 40, 70–71, 128, 129–31, 132–36, 137, 139–45, 147–52, 166, 172
Rio Grande Compact, 140, 143, 149, 150
Rio Grande Project, 140, 141, 145, 148
Rodeo, NM, 15
Ruscavage-Barz, Samantha, 51
Rust, Philip, 116–17

sacrifice zone, 56
San Augustin Plains (also, Plains of San Agustin), 112, 122–27
Sandia Mountains, 67, 101–2, 113–14, 116–18, 145
San Juan Basin, 43, 47–62
San Juan-Chama Project, 11
San Juan Citizens Alliance, 55, 58
San Juan River, 11, 49, 50, 77
Santa Clara Creek, 97–98, 135
Santa Clara, Pueblo of, 88, 97–98, 134–35
Santos, Fernanda, 108–9
Schmitt, Harrison, 40
Silver Fire, 40, 67–68
silvery minnow, Rio Grande, 29, 70, 169
snow: "snow drought", 4; reduced snowfall, impact on Colorado River, 10, 131–32, 133; reduced snowfall, impact on Rio Grande, 131–32, 133–35, 140, 141, 157–58; snow-pack, 101–3, 121, 129, 146, 160
Solomon Islands, 22
Southwest Jemez Mountains Resilient Landscapes and Collaborative Forest